JUST DO THIS:

A SIMPLER WAY
TO SUCCEED IN I.T.

D.J. ESHELMAN

Just Do This: A Simpler Way to Succeed in I.T.

Copyright © 2020 by D.J. Eshelman

ISBNs for Just Do THIS-

Hardcover:	978-1-952105-10-4
Paperback:	978-1-952105-07-4
eBooks:	978-1-952105-11-1
Audiobook:	978-1-952105-08-1

Kindle Series-

Part 1: 978-1-952105-03-6
Part 2: 978-1-952105-04-3
Part 3: 978-1-952105-05-0
Part 4: 978-1-952105-06-7

Cover creation provided by Amir Hegazi via 99Designs.com

Editing support provided by Lorraine Reguly

Version 1.4.0

Dedication

This book is dedicated to those who dream of having the opportunity to succeed in Information Technology, without regard to color of skin, gender, wealth, or how they grew up. This book is intended to help level the playing field.

You're about to learn what made the difference for a kid growing up poor with no running water or electricity to earning a consistent six-figure income and traveling the world as an in-demand consultant and speaker. I wrote the book I wish I had access to 20 years ago in the hopes that you will be able to accelerate your career for a better life balance… and that you will in turn pay it forward.

If you're willing to do the work, this book is for you.

This book is also dedicated to my mother, who taught me the value of work and dedication, and to my wife, who is the most encouraging voice in my daily life. Thank you for encouraging me to *Just Do This*!

Good news!

As part of your book purchase, you are eligible for exclusive bonuses, including the companion electronic course (worth over $199), the documentation tools I use in Assessments and Health Checks, PDF diagrams from the book (in color), printable PDF workbooks, and the audio version of the book.

I want to see you succeed and I'm excited to provide these additional tools valued at over $500 USD to make sure you are successful. As new opportunities emerge, I will email those out to those who register.
I'm on a mission to create $100,000,000 in raises, bonuses, and better career opportunities.

It starts right here. *Just Do This*!

Register at
https://it.justdothis.net/mybonus

Praise for Just Do THIS

"In this book, D.J. gives away the secrets on how to succeed in IT in simple steps that anyone can learn from, no matter if you are new to IT or an IT Pro. Beautifully written and easily understandable. It is highly recommended reading for everyone in IT today!"

- Douglas A Brown
DABCC Radio / DABCC.com
Vice President of Community, IGEL

"D.J.'s Just Do This book is a must read for anyone in the IT services industry. This means everyone, since even if you are internal IT, you are providing a service to your organization. D.J.'s experience will guide you through how to operate and why… regardless of the specific technology area you focus. Read it… Just Do IT… I mean, THIS!"

-Benjamin Crill
www.icrill.com

"D.J. Eshelman provides practical advice and real world examples to be successful in your IT career. I have personally followed his methodology and have successfully implemented new technologies for several customers."

-Abraham Brown

"Of all the technical books I've read over my career, I have to say Just Do This has a fresh and enjoyable style. The book is full of 'don't be a chucklehead' advice with a personal touch, clear and concise suggestions mixed with anecdotes and experiences that shaped the author's methodology. While specifics are given regarding IT projects, his methodology can be applied anywhere. An educational read, I recommend the book for anyone looking to reinvigorate their career and step up their professional game."

-Steve Wightman

Forward

"For over twenty years now, I have been working in the IT space, and during that time, I have always had at least some focus in the end user computing experience side of the space. Let's face it, that is the entire reason for IT at any business, to give your employees a great working experience so they can best service the businesses customers. If it were not for that purpose, IT would not be where it is at today, not even close. During my time I have seen many different project methods come and go, all with varying degrees of success. ***After reading this book, I have been able to work with two large companies and put this method into practice. The results are amazing.***

My first experience in using D.J.'s methods was with a healthcare company in Ohio, I had no idea how it was going to go work, especially since it was my first meeting with a new Account Manager on my team (I am an SE.) After we got through the introductions, and small talk, I wanted to get to step number one, right away UNDERSTAND. During the car ride from Pittsburgh with the account manager, we talked about getting to know why we were there, if we did not UNDERSTAND, we would not be able to add any value. Even more so, if we did not UNDERSTAND, we may have ended up going back time and again for no reason, when we should have found out in the first meeting, we could not add any value. Knowing this all important first step, was our main goal of the meeting; and it worked.

Once we knew what our role was, and why we were being evaluated for their work from home initiative, we were able to develop steps and goals. We now had ownership and responsibilities, meaning we could clearly focus. During this phase, we acted as trusted advisors or as D.J. calls us – sounding boards, and we were able to keep our focus on the *why*, stay on task, and be mindful of other agendas. Several times other agendas tried to get pushed, but since we both knew the *why*, we were able to keep going back to that and showing the progress we had already made. When we did this, it instantly stopped those outside influences.

In a dramatically short period of time, our technology had been chosen to be tested, we had established success criteria, timeframes, and had a specific focus. It was now time to move onto the next step, PLAN.

When it came time to develop the actual testing plan, I was feeling confident as I had just read the second part of the book – PLAN. So right away, I knew I needed to keep it simple and my account manager and myself, both had to have a different plan. This was different than any other deal or project which I had worked on before, so there was some anxiety, but it was the good kind. My plan was to take the success criteria and develop an implementation and testing plan around the work we had already done and not allow and scope creep. For the account manager, he needed to get to work with the project sponsors and owners to learn exactly how and when we fit into the buying cycle. Knowing our tasks ahead of time and having everything documented, proved to be a monumental success. The PLAN was simple, easy to follow and repeatable on the technical side. Combined with the work from the account manager, our technology was purchased for a small user acceptance pilot. This is where step three started – CHANGE.

Now that we were in the CHANGE phase, we brought in our Advanced Services Team, who documented everything and worked with the IT team at the Healthcare system to hone everything in for the optimal end user experience. Nothing but tiny micro changes were made and tested, one at a time, and each one was documented. Knowledge was transferred and taught every step of the way. **Following these principles, the relationship built between our two companies turned into a partnership**. Once all of the success criteria was met and the UAT pilot was done. It was time for us to move into the fourth step – MAINTAIN.

For me, the MAINTAIN part was a whole new concept, and something I was missing from my toolbox, for my entire career.

Once I understood the method behind MAINTAIN, I was convinced this was going to advance my career and the career of others.

When we had our success call with the Healthcare company in Ohio, everything had gone so well, they had leftover consulting hours, the account manager and I were faced with a decision. We could turn those hours into more product, or we could apply them to a resource to help them maintain their new platform and let them know when it was time to go back to UNDERSTAND, PLAN and CHANGE. We choose the second option, and the *healthcare company could not be happier*. Now they have an expert to rely on, to help them maintain, and update their new platform, document the processes, and train new people coming on board. *This also works out great for my company because it will keep the help desk tickets down and help keep me free to put this method into play at more companies and make more satisfied customers*."

-Douglas DeCamp
Enterprise Presales Engineer
IGEL

Table of Contents

Introduction—WHY

"When you know your WHY, you have options on what your WHAT can be... When you know your WHY, your WHAT has more impact because you are walking in or towards your purpose."

—Michael Jr., Comedian and Motivational Speaker

Let me begin by saying I'm not trying to present to you a definitive work, a standard by which all others are judged, or the dreaded "best practice" speech. This also isn't some pie-in-the-sky notion of the newest and most complicated method to be "Agile" or build an overly complicated and risky method of doing DevOps which will revolutionize the industry FOR-EV-ER.

Instead, I am attempting to present to you a distilled, finely honed, and simplified method of improving your professional life, your professional services practice, or your sales acumen. It can be utilized in any number of other applications I have found for what follows. It, by no means started as my idea. In fact, I credit Citrix Consulting with the vast majority of the concepts I am putting forward for you. I don't know for sure, but I'm reasonably certain they didn't invent it either. Like so many things in Information Technology, the process has evolved and improved over time. I'll say, with confidence, I'm growing it as well. So, if I'm stealing, it certainly isn't on purpose. My research could not come up with definitive prior works. As it turns out, this is largely another reason for why I created this resource for you—

because someone should be teaching what REALLY works!

However, also like so many things in Information Technology, the process can sometimes be so complicated as to be bypassed, changed, or ignored altogether. It is the latter that concerns me the most and was the singular inspiration for me writing this book.

What I can tell you is in the years before I worked for a large worldwide consulting organization, I was already noticing that my clients were very often frustrated. Tasks that should have been simple became complicated. They often took risks from the pressure to be more "Agile," for which they weren't ready. Very little was documented in the environments in which I found myself working, and all too frequently, I would be tasked with "taking over" for someone who could not articulate where they were in a process because… they had no process. They just showed up to work, broke some things, fixed others, and were quite often either fired or sent to work in the Spice Mines of the Service Desk.

During my time with this, and other consulting organizations (including my own), I learned how and had the opportunity to help them grow various processes of their methodology. I taught their staff what I'd discovered about how to have proper documentation processes— methods which have continued improving since I left. I learned, firsthand, the difference between a professional consulting group that lived up to the value of their hourly rate just from having a better process than the companies they were serving, and the groups who did not.

As both a Resident and an Independent Consultant & Architect, I have helped several other companies hone or start their own services methodology. One, in particular, had a Professional Services process which involved over 12 steps! They got an "A" for effort but a solid "C" for execution, most of the time.

I learned to **keep it simple**, a concept you'll see throughout this book. As a part of being on various IT teams, either full-time or from contracts, I learned the value of a daily mindset of what I'm going to teach you. Change Control meetings went smoothly, and, most importantly, there were fewer outages caused by changes being made. We spotted problems earlier and reduced costs.

Contrast that with a friend who has his team obsessed with the "new" way of working—an overtly complicated "DevOps" inspired methodology—which has tripled the amount of meetings, doubled the amount of staff effort, and caused several late nights of working... all to get a project "ready" faster.

We probably shouldn't even talk about what happened when it all went horribly wrong, but I hate to see those "resume-generating opportunities" happen.

I hate to see things go badly like that for the customer. I really do.

Equally dangerous is the reality for a misguided acquaintance who runs a small Managed Services Provider and cannot seem to escape work. He and his staff often work around the clock and they are never really able to relax. They are always fearful of what is around the corner, and typically serve the loudest complaint because they operate with no real methodology at all. More or less, projects are swept aside until there is a loud complaint about it which happens to be louder than the trouble complaints (which should go into a ticket system that they don't even have). A simple methodology, by which everyone could realistically abide, would easily boost their business output by 50% or more, make for happier customers, and boast a staff that doesn't "rage quit" (as they have dealt with so often).

That doesn't even scratch the surface for the crushing reality of about 80% of the companies for which I worked. Their stories always seemed to be the same—they have a small team that is already stretched beyond their capacity, and the team members rarely have the knowledge or experience to handle the technology with which they are tasked.

It isn't just that they don't appreciate the true nature of work, it's that they rely on contractors to do so— contractors which are usually there for a short period of time to accomplish a single goal. This is not bad in and of itself; I would argue a company that doesn't rely on contractors in some capacity probably isn't growing. The issue is, they have staff who don't want to learn complicated methods. At the end of the evening, they want to go home to their families. Adding contractors or throwing more staff at the problem doesn't fix it. In fact, sometimes adding a contractor makes the staff groan because their eight-hour day just became an eleven-hour one.

And that is the bottom line: WHY a methodology is essential… is as important as WHAT it is.

To me, there are several reasons to embrace a proper methodology, primarily which are:

- Reduce risks
- Reduce costs
- Leave things better than you found them
- Avoid resume/CV-generating opportunities
- Advance your career
- Take vacations
- Never worry about your evenings again

I'll discuss the *why* of developing a methodology in much more detail as we go along.

Let's begin with the end in mind and start with what a methodology is!

A Professional Services Methodology is an iterative process that defines the tone and purpose of an interaction to generate a specific desired result.

To best accomplish this, the methodology I am starting you with has four phases:

1. *Understand*
2. *Plan*
3. *Change*
4. *Maintain*

I'm willing to bet, if you work in Information Technology at all, you already are familiar with these concepts, but it's possible you never gave them a name. Most importantly, don't miss the keyword of "iterative" in the definition of the process. Iterative, in this case, means the intention is for each phase is to feed into the next, allowing for the possibility that it may need to come back with corrections.

Iterative, in that you always expect the cycle to restart, but knowing each goal must be met before you continue.

Put simply:

After you *UNDERSTAND*, you *PLAN* what you will then *CHANGE*, with the goal to *MAINTAIN*. Then, in maintaining, you will also discover new needs or requirements, which leads you back into *Understand*, *Plan*, *Change*, and back to *Maintain*.

Please enjoy learning this process. You will emerge fully equipped to take your career to a whole new level of enjoyment, and perhaps even bring your entire company to a better way of operating.

—D.J. Eshelman

Chapter 1: Concerning Methodology

"If you aim at nothing, you will hit it every time."

—Zig Ziglar

The methodology I have developed has four simple but iterative steps:

1. *Understand*
2. *Plan*
3. *Change*
4. *Maintain*

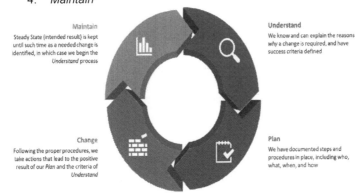

Maintain
Steady State (intended result) is kept until such time as a needed change is identified, in which case we begin the *Understand* process

Understand
We know and can explain the reasons *why* a change is required, and have success criteria defined

Change
Following the proper procedures, we take actions that lead to the positive result of our *Plan* and the criteria of *Understand*

Plan
We have documented steps and procedures in place, including who, what, when, and how

At times, I have wondered if overly simple is good or bad, or if there are steps missed in keeping things simple. There may seem to be, but my axiom is this: KEEP IT

SIMPLE. If it is too complicated, the possibility of things that can be lost or skipped over increases.

I wanted to put forth a methodology that can be used in multiple phases and industries. This process works within sales, consulting, engineering, software development, and even process-heavy DevOps models.

To make sure I was not alone in my thinking in this, I performed a brief search on Google for Professional Services Methodology. The results were overwhelming in number, but the most important thing I discovered was, I am certainly not alone in my way of thinking.

Almost every graphic demonstrated an iterative process of between four and fourteen steps, usually four or five steps.

Almost all had names that were similar, such as:
- Strategy, Review, and Planning
- Discovery, Analysis, and Design
- Verification and Action Plan
- Implementation and Execution
- Quality Assurance and Measurement

One company I found in my search is one I have to admit to being a little jealous of: a company called N8 Identity.

Their methodology was listed as Strategize, Visualize, Realize, and Operationalize. That is "brilliant*ized*" wordplay!

Who inspired me the most, however, is Citrix Consulting. They use a very well-known methodology of:

- Define
- Assess
- Design
- Deploy
- Monitor

You may have seen this previously as:

- Analysis
- Design
- Build/Test
- Rollout

It is also possible you have seen the Microsoft Services Framework, which is similar:

- Envision
- Plan
- Build
- Stabilize
- Deploy

So, why our four steps? And why such simplistic names?

The answer is because I want to KEEP I.T. SIMPLE BY KEEPING IT SIMPLE!

A methodology must be:

Universal. The entire team (management, contractors, and everyone else) must understand WHY. This is the key to having everyone using the same methods. This keeps goals achievable and recognizable by everyone in the organization—from Sales to the Service Desk to Management. Everyone must understand it.

Goal-Oriented. Each step indicates the goal, not the effort by which the goal is achieved. It is how we think, and the method should reflect it. After all, the efforts to achieve a goal will change based on the need!

Easy to Remember. You should be able to quickly recall and communicate to others where you are in the process. I used to refer to "*Change*" as "Deploy" just as I used to refer to "*Plan*" as "Design," but then I found just having the first letter the same was causing confusion. Moreover, what if the task at hand didn't imply a design but a quick change? The notion of an entire design was exhausting when only a plan was needed! So, I changed the name. Doing so seriously delayed the project. You'll still find me slipping old names in every now and again, but if you're new, I am confident these simplified descriptions for each phase will stick with you.

Simply Understandable. It must make sense! The words I chose reflect this. You might be looking at them and wondering why a whole book needs to be written around them. If that's the case, the first part of my mission has already been accomplished! Contrast this with complicated terms in ITIL, Agile, or other methods and you'll immediately understand why I wanted to create a system which doesn't require a certification or degree for everyone to implement. Everyone is invested as a result.

Repeatable. The process should never be closed off. You should always present the opportunity to improve. The common saying in the Open Source community is you would never buy a car with the hood welded shut, so the process should be mindful of repeating itself, on purpose—not because it failed, but because it succeeded!

Part of your Daily Process. You must be able to live in this methodology on a daily basis, especially if you have multiple projects going on. I encourage project managers to include an awareness of not only where a project is in the process, but where it is going next. They can then best communicate with management and project members alike, the tasks needed for the day, even if they are working on multiple projects at once! Ideally, this means even your daily emails should include where you are in the steps for a particular call to action or status update.

Adaptable. It must be relevant to multiple situations. If the sales people and engineers are in line with the phase their customer is currently at with a project, they can both adapt their strategy accordingly.

In my same search, I found several companies that displayed complicated methodologies on their websites. Some were so complicated I stopped reading after two of fourteen steps because I had lost interest—and I was searching for it, specifically! If you cannot simply and effectively explain your purpose, you will lose your audience and possibly even your own focus.

I also wanted to use words that can describe and pertain to more than just a single project. I literally want this to be a lifestyle for the technology world. The more we can speak the same language, the more we can quickly succeed together.

So, keep it simple! You don't have to use my exact words, but the key is to use SOMETHING, and to use it consistently in everything you do! I researched more than 10 IT Services companies, and common threads emerged. It is those common threads on which we will focus. What I found to work in companies in the long term are simple methodologies that everyone could understand. More steps and more complicated descriptions led to siloes in adoption of a methodology. My encouragement would be to keep it simple and *JUST DO THIS*.

Football and Methodology

In the United States, many of us are borderline, if not completely obsessed, with football (not soccer, but follow me because this is going to work for both sports). One particular obsession is fantasy football. Fantasy football is aptly named because it is a concept of putting a bunch of the best players you can find together on a fictional "team" and seeing how well the stats work out.

Let us go a step further and pretend this could actually happen in real life.

What if we could get the absolute best players in the game all on one "super team" and let them play everyone else? Now, we know these players have talent, knowledge, and the ability to perform at a peak level of what humans can do. So, we are going to let them just play the game the way they think it should be played because, they are the experts, after all.

Let's put them up against a team of rookies. That team will run drills and learn plays. Above all, they will be coached in such a way that there is a universal coverage of all duties by the players—more than having only the best at each position.

Who do you think would win?

I would put my money on the team of rookies with a good coach over the team of high-performing experts any day of the week. This is because even the most talented individuals—if not unified in a plan of action which is repeatable, predictable and well-defined—will stumble. A team not possessing the ability for multiple members to be able to perform the tasks of others when needed will be outmatched by preparation and methodology. Egos will take the place of teamwork. This is human nature.

Yet a shocking number of IT teams and consultants perform their jobs as a collection of experts, tripping over each other, arguing, and often taking steps to sabotage each other simply because they don't have a system in place that lets success be measured by collective results rather than individuals not achieving. When things go wrong, there is a race to find someone to blame. When things are missed, lack of experience is blamed, at times, but quite often, it is the individual who gets fired for missing the mark. Does that sound familiar to you?

Now, what if you had a team of high-performing individuals who submit to a process of coaching, discipline, and willingness to go after progress down the field instead of looking for the "big play" that will boost their stats? How do you think that team will do? I believe they would win the Super Bowl!

I grew up watching a particular US football team called the Pittsburgh Steelers. Up until 2019, they had won more Super Bowls than anyone. That team, with a powerful legacy, has only had three coaches since 1969. Those three coaches have some of the most amazing winning statistics from teams who produce fairly few Pro Bowl participants. You would think that the team would be dominating either the Super Bowl or Pro Bowl every year, right? Yet, aside from obvious standout stars, it is how the team performed as a whole that makes them contenders year after year, getting to the playoffs more consistently than their peers.

I tell people all the time, the Denver Broncos weren't my team even though I lived in Colorado, because I did not agree with the way they were coached. Their proclivity towards seeking talent over guidance as a cohesive unit shows in the sheer number of coaches they have had, and the number of player trades made. My allegiance was far outside of my state simply because of how I have grown to admire a process being more successful than individual "experts."

Are the Steelers perfect? Heavens, no. With pride comes the occasional fall, and I have been known to go hoarse from yelling at my own team to stop making stupid mistakes and getting penalties which cost them games. Yet, I remain a proud member of Steeler Nation, because it is about the team working together (more than the talent of an individual). One need only look at what was often called "The Steel Curtain" and some of the cohesive plays that took place during then, and other eras, which provides endless entertainment and excitement. Even games that were lost still felt like wins when there were plays you end up talking about for decades.

It is with that overall attitude I approach this process. I believe each one of you, regardless of your level of experience, can grow into a successful career with less stress.

Terms Used

A few terms and the reasons why they are specifically used in this book need to be defined.

Leading Practices. I credit my friend, Nick Rintalan, from Citrix Consulting, for teaching me this. "Best practice" is a statement that is often thrown around in Information Technology that really should not be, in the vast majority of cases. Very rarely are you in a position to make a value judgment as to what is "best" in a situation. By claiming something is a "best practice," you are often unintentionally stating your recommendation is 100% true for the situation you are speaking to. Using "leading practice" instead of "best practice" gives room for the reader to understand, even though every situation can be different, the vast majority of companies do it a certain way. That should be what people start or "lead" with when the question comes up. The term "leading practice" also limits your liability in making a bad recommendation.

Scope. In every engagement, whether it is for your employer or someone for which you are consulting, it is crucial you understand the areas in which you are engaged (and not engaged), what the goals are, and what needs to be avoided. This is called "scope" because, much like when you look through a telescope, you are looking closely at certain items but limiting your vision around yourself or other things.

When you are reading or writing a scope, think in this way:

1. What items do I need access to in order to be successful?
2. What things would cause risk in the project? What things might be asked of me while I am in the process that would cause risk or delay my ability to complete the project?

The scope is there to focus you, and the people you are engaged with, to ensure that you are all in agreement of what is to be done.

Scope of Work. Typically, a non-binding document in which two parties are working to define the areas of a Scope that will specifically apply to a team or team member. The Scope of Work is often used to determine agreed-upon Statement of Work documents, which define what will be done, when, and by whom.

Test. Possibly the most important word an Information Technology professional can utter is the word "test." I say this because no recommendation, design, plan, or process is perfect. The user should always be protected from unstable changes in an environment. This is done by testing or validating a change, prior to it being applied in production. Quite often, an organization will need to have a dedicated mirror environment of what is in use to properly test. We will discuss testing in much greater detail in the third step: *Change*.

Production. Simply put, this is where users are interacting with live data. Production is the normal, day-to-day operation, and it should be your goal not to disrupt it. In fact, I would argue that disrupting production can be one of the most blood-pressure-spiking experiences a person can have. That feeling in your gut that you have done something awful is because, if you have disrupted production and could have prevented it, you really did do something awful. So, protect production. Never ever test in production if you want to be considered a professional.

Stakeholders. Those who are ultimately responsible for assuring the systems are properly operational are the stakeholders in a project, who are very often IT Managers, Directors, Vice Presidents, or even the Chief Technology/Information Officers, depending on the company. People are typically designated before a project engages.

Project Sponsor. Often, a company needs to define who will advocate for a project, protect its budget, and assure to a company's leadership that the project has merit. They may sometimes be the project stakeholders, but are very often the people in Finance or a part of a specific project team you are serving.

Audience. I use this term in the book to reference the people who will be reading your documents, both now and in the future. You may have an immediate audience of people in a conference room, but you may find at the end of a presentation they will ask you for the slide deck. When that happens, your audience is no longer the people in the room—your audience becomes the people viewing the slide deck.

Spoiler alert: you may never encounter those who view your document. This is the same with documentation. Your audience is who will read the document, both now, and perhaps years from now. Being aware of this and forming your content to match is a differentiating factor for a new professional.

Success Criteria. This is a mutually agreed-upon set of intended results which will be used during all stages of testing to determine if a project is successful or not.

Deliverable. This term usually describes a certain aspect of an intended result of a project. For example, deliverables may be listed as "An optimized desktop operating system image," or "Design Documentation," etc. Documentation deliverables are most common, so for the sake of this book, we will mostly be referring to documents as deliverables.

Layer. A document, like the ones we will be targeting to produce, should separate sections to differentiate focus areas called layers. Like a cake often consists of multiple layers, so can be described of a computing environment.

Layers

To maintain consistency across all types of document, discussions, and presentations, it is best to standardize an approach of describing "layers" for each technology focus area. You should feel free to make this your own, but I'm going to describe the standard that is used more or less by Microsoft, Citrix, and VMware to describe end-user environments. If you run DevOps in your software development process, for example, you may need to modify these layers, and I'm sure some will not be required.

Business Layer. This describes the "why" of an overall environment. What does the environment do to support the business as a whole? What justifications exist for its current or intended configuration? For example, a Citrix XenDesktop design may contain the environment described in the Business Layer as being "A Virtual Workspace in which users can work consistently from most locations and nearly any device, thus lowering operational IT costs by standardizing a method of access, increasing security by keeping data inside the Corporate Cloud, and allowing work from any location with Bring Your Own Device policies, which reduce operating and capital expenses year after year." Regardless of technology, if purchasing something doesn't help the business, what is the point of buying it?

User/Subscriber Layer. This section describes the ways in which users are interacting with the system today or intend to interact with future systems. It will typically include a statement of the user's needs, a description of their methods of working, the hardware supported or intended to be supported, and the needs for identity and security management for the users. Often, any compliance need is described here. In our Citrix example, the User layer would include an overview of each grouping of users (use case), along with how many are in each group. This overview would include for each use case where and when users are typically logged in, what software packages they require, what persona management is required, and any compliance needs are present per user group. The methods by which the Citrix Receiver is maintained on endpoint devices is often described as an overall idea of how users maintain their systems. This is at a basic level; much more information is typically contained here.

Access Layer. This section describes the methods by which users are given access to interact with the system (as previously described). This will include descriptions of the current networking capabilities—at client, server, and WAN/LAN scenarios. It will describe how subscribers, both internally and externally, will be granted access. For our Citrix example, you may describe how all users will be connecting the Citrix Receiver (now called Workspace App) via https (port 443) to a NetScaler Gateway (Citrix Gateway, as it's now called) to proxy the ICA traffic. Internal connections over MPLS will connect to the Local Access Website (e.g., https://access.website.local). External users will be using a portal which requires a multi-factor authentication at the Access Website (e.g., access.website.com) that uses GSLB location awareness to determine which data center to authenticate against. External access is X Mbit/second, which will accommodate the X user workload estimated at X kbit/second per active connection. Requirements for servers controlling access and hardware required are outlined.

Resource Layer. Often referred to as the Application Layer in some designs, this section describes the applications themselves, whether they are located on a Windows or Linux desktop, for example, or in Web SaaS applications. Resources being consumed are mapped to the Users or Subscribers (noted previously), and the needs outlined are described in terms of how many resources will be required to accommodate them. Typically, a lot of tables are used in these sections, especially for a Workspace/VDI Transformation project. Do not worry about the physical requirements required for the Resource Layer to be accommodated in this topic.

This topic is merely to count the requirements that will be met in the Platform Layer. This section identifies the needs and typically has a table at the end which summarizes the information. In the XenDesktop Design example, this section would include a listing of all the master images required, the virtual configuration required, the number required, and the software to be installed to each image. Some configuration details, such as optimizations and load management, may also be described here, although this often falls into the "policy" category described in the Control Layer.

Control Layer. Often split into multiple parts, the Control Layer section describes the methods by which the user identity, security, policy, and other needs are controlled. In a XenDesktop environment, this may include things like Active Directory structure and policies, XenDesktop controllers, policies, persona management, SQL databases, image management, and more. As you can guess, this section will be very large, often with links to appendices detailing design decisions made for Group Policy and Citrix Policy settings. Hypervisor and Cloud connections will be outlined and mapped to resources, as appropriate. In a design document, the sizing and HA requirements for the components will be outlined and summarized at the end.

Platform Layer (often called the Cloud, Compute, or Hardware Layer). Now that we know all the resources we need, where do we put them? This is where the conversation has gotten really interesting in the last few years, with the proliferation of cloud-based workloads combined with traditional data center workloads. The Compute layer is where we ensure that we are meeting the CPU, memory, disk and networking requirements for Access, Resource, and Control components. Physical and Virtual Cloud locations are discussed, as is the exact hardware configuration, when appropriate. For example, sizing a XenDesktop VDI environment requires thought as to how to best optimize the physical hardware and "right size" for the CPU and memory. If purchases have not been made, recommendations or requirements would be outlined here to best utilize the hardware. In many cases, load testing results are described here, which outline how resources will best be spread across the cloud instances (and yes, I consider "cloud" to simply mean "a data center" or "someone else's computer").

Security Layer. Sometimes, combined into the Operational Layer, this section describes how user data is kept safe. Compliance such as HIPAA, PCI, SOX, etc., are discussed. Policies for anti-malware and data security are outlined.

Operational Layer. The key to the *Maintain* part of the methodology is the Operational layer. As a matter of fact, the only reason I don't just call it the "Maintain Layer" is because so many other industries use the Operational terminology. There are both human and policy elements to keeping all of this together, and the Operational layer section describes the *who*, *when*, and *where* of the administrators and staff maintaining the environment. Service desk permissions and escalation roles are defined. Disaster Recovery and Business Continuity are discussed. Update and testing procedures are listed. How the environment will be monitored and methods of proactive and reactive monitoring are described.

Also, of note is, this list changes every now and again as technology changes. In fact, I made changes to the original list I made in 2017, to a newer and updated list in 2019, because of changes in Cloud computing.

Example Projects

We will use several example projects in our discussions to illustrate exactly how the methodology works in each. By no means is this a definitive list. As a matter of fact, they are simply the projects I know best in my professional life.

Check the Methodology Book blog page for more examples, found at https://it.justdothis.net/examples.

Workspace Transformation

Perhaps the most complex projects I have been involved with, the Workspace Transformation (sometimes called Desktop or Application Virtualization, or similar names), involves taking existing or new use cases, often with existing PCs or servers, and transforming them into a virtual or cloud-based workspace. Often, the goal of such projects is to bring all data into the data center and restrict the outbound data to replace things like VPN connections. Leadership may want to use thin clients or support Bring Your Own Device (BYOD). In that case, the Workspace Transformation project may involve data collection from multiple PCs, a detailed assessment, and a Use Case Assessment. The *Plan* phase will certainly involve lengthy Design Decision meetings, whiteboarding sessions, diagrams, and an examination of several associated technologies, across each layer.

Workspace Transformation projects have a very lengthy *Change* phase, which involves the coordination of building components and validation testing, all which typically involve a great deal of user interfacing and education. These projects quite often represent a whole new way of maintaining applications and user personas within the environment. As such, the *Maintain* phase should be part of the planning and documentation process.

What is especially unique about the Workspace Transformation process is that often there is a sales cycle involved that should follow the same methodology. It doesn't matter if this is for an external team selling a project, or an internal team "selling" management on dedicating resources for the project. Without a proper *Understand, Plan, Change, Maintain* cycle involved in the sales or convincing process, it runs the risk of just being another "pet project" that will likely be ended mid-stream or not be done to its full potential.

Health Check

This brief assessment focuses on a specific technology area to determine if a previously deployed environment is running to expectations and scores the appropriate risk areas. Health Checks can sometimes be called upon when the troubleshooting of a problem is not going well and management would like to rule out deviations from leading practices. The goal of a Health Check is to determine if additional projects, products, or even staff are needed. Successful Health Checks are always tactical and are rarely conducted for more than a single week. Though, labeled this on occasion, they are also not an Infrastructure Assessment.

Infrastructure Assessment

A deeper dive into a specific technology, with a specific awareness of the Business Layer requirements of a deployment, is called an Infrastructure Assessment, which provides a combination of both strategic and tactical suggestions. The Infrastructure Assessment typically involves an examination of the leading practices in multiple layered elements because one system often affects others. The Infrastructure Assessment process often takes between one and four weeks as a detailed analysis is performed. A peer review is a crucial step in conducting these kinds of assessments, and I always recommend having a qualified and certified technology professional review your documentation before it goes to management. The documents are quite often very lengthy, often exceeding 50 pages. Detail is important, but guiding towards strategic and tactical suggestions is the primary goal. This type of project will often include taking remediation steps during the *Change* phase.

Windows Updates

You may be surprised to see Windows Updates on my list, but I feel it is always important to practice this methodology. This is just as true of a weekly event as it is with one that occurs every three years. The more you take the correct procedures into account, the more you will find the work becomes easier and the risks so low, you are finally comfortable letting others take on tasks because your system is working so well.

Our scenario is going to be a common one because Microsoft releases updates on a regular basis. "Patch Tuesday" is a common phrase to hear around the office. The issue, if you aren't aware, is that often patches to fix one issue can negatively affect others. Microsoft expects you are testing properly and therefore releases patches with a regular cadence that some refer to as "public beta tests."

Although the track record is actually very good, I always recommend testing, because cleaning up or recovering from a bad patch can be more expensive than the flaw it was intended to fix! So, *Understand* the need addressed, where appropriate, *Plan* a method and time to deploy, and schedule the *Change*, then monitor the results (and *Maintain* them when they are good). It is simple enough and it follows the methodology well.

Part 1: UNDERSTAND

Chapter 2: The *Understand* Phase

"Begin with the end in mind."

—Stephen R. Covey

Getting to Know the *Understand* Phase

There is a bit of irony in the order of my methodology in which people are exposed, because it tends to happen in reverse. Most start in a helping, monitoring, or administrative capacity—the *Maintain* phase is where their job is focused. Engineers, on the other hand, tend to be focused on the *Change* phase for their daily tasks. Architects and Consultants are typically tasked with the *Plan* phase and may also need to be the driving force in the *Understand* phase. So, your exposure to a proper methodology happens naturally, but in reverse.

Much like a soldier following orders, however, if you do not have a good grasp of the process, you are more likely to fail. This concept is talked about in the excellent book *Extreme Ownership,* by Jocko Willink and Leif Babin (see https://it.justdothis.net/bl#1). Teams that do not understand the mission cannot take ownership of it.

Therefore, I encourage you to pause where you are in your career and consider what is happening outside of your area of focus. If you are in a role that is focused on either of these first three phases of the methodology, I encourage you to consider other roles that will fulfill the vision and support it. Are you creating systems that can actually be maintained? Can your vision be maintained? Or are you taking enough actions to meet long-term needs so that another team won't be "stuck with the bill," so to speak? In other words, I want you to begin with the end in mind. With the vision of having a well-maintained and well-designed system or result, the first phase in our methodology approach is *Understand*.

Understand, sometimes called Assess in I.T. Services, is the process of observing what the current needs, configurations, systems, and risks are, and reporting them in a standardized way which leads to the next step of the process: *Plan*. The result should always be a conveyance of the fact that you understand and can describe something properly, which can take many forms, depending on the overall initiative and what deliverables are desired. Nonetheless, in all cases, *Understand* focuses on what is needed. In some cases, the *Understand* phase makes suggestions on how to improve or focus.

For example, say a company would like to do an overall Infrastructure Assessment to identify any deviations from leading practices and any risks that exist in the environment. The deliverables for this type of project would consist of a focused presentation to upper management and a detailed document describing the findings and suggested courses of action.

An important aspect of this kind of undertaking is fully understanding your audience. In considering our audience, we must think outside of a single scope of maybe a group of engineers or Information Technology staff.

We must make recommendations which address the company's interests as a whole. This means being intentional about the tasks and data presented (and not presented), as well as being consistent in your message. If you are primarily going to be addressing a technical audience, keeping the findings as straightforward as possible is best. If you will be addressing a primarily management-focused audience, you will want to shift your content to explain things in a very conversational way (which we will discuss soon, in another chapter). In either case, it is of crucial importance to make sure you get the same message across where it matters most.

Tools and scripts are quite often used during assessments. However, it is very important you understand that not every assessment will have a tool to make things easy. A lot of the time, you will be required to get into the systems and manually document their configurations. Quite often, this will mean having a good idea of the overall project scope.

In the event you are assessing as a precursor to a design, you will be gathering information but may not have an existing design to reference. In those kinds of projects, your goal is to define the success criteria and requirements which will feed into the design in the *Plan* phase.

From here, we are going to build out a standardized structure of how to document observations made in the *Understand* phase. The document structure introduced in the *Understand* phase will be followed, more or less, for future documents, so it will be the largest topic discussed herein.

I would recommend having your notebook and pen handy—or even better, get ready with tools such as your favorite note-taking application, word processor, and/or spreadsheet applications. Most people (like me) learn more when engaging multiple senses at the same time.

Business Needs Assessment

Contrary to what may end up in practice a lot of the time, I believe all IT processes should begin with the vital element of understanding of the business. I will refer to this later as the Business Layer.

This element guides everything about what the project is working towards, as a whole.

Taking the time to understand the company business is often called gaining acumen.

Assume, for a moment, that you intend to build a multi-story building. Now, imagine you show up to the job site and just start doing what you know how to do—putting up steel, concrete, and wood until you have the ground floor built. You move onto the next, and the next, and when you get to the fourth floor, you begin to notice something is wrong. Part of the building is no longer level.

You review your measurements. Everything you have built is perfectly proportional and even measures properly, yet every floor is dipping downward from the rest of the floor at one particular point. *Not my problem*, you think to yourself. *I'm doing everything by the book here. This is my scope of work.*

You probably know where I'm going with this already, but let me say it: you didn't consider a foundation! Perhaps, in your previous experiences, someone else always took care of it for you. Perhaps you missed it. But, without a firm foundation, the building will eventually crumble, even if everything that follows is 100% flawless.

Conducting an Assessment without first understanding the needs of the business you are serving will fail in the same way.

Ask yourself, *Am I giving an unfounded opinion or a relevant analysis that I would pay for myself?* If you haven't taken the time to understand WHAT the business is doing, and WHY the particular project you are involved with is important to accomplish business goals, you are not providing value.

I am consistently amazed by the number of times I see Assessments given (formally or informally) that have not taken the time to understand WHY a business has a need. Nearly 100% of the time this happens, the Assessment ends up thrown out. It is a waste of time, read by few, and heeded by even fewer. So, even if the project is a government job to simply eat up some budget money, take this seriously. As a taxpayer, you should take it seriously anyway!

Let me give you an example of why this is so important.

A company has engaged you for a project to migrate their on-premises email "to the cloud." *No problem*, you think. *I have a service in mind which can do that and save the company thousands of dollars a day!* You begin the process of looking into how much email data there is and find there are under 2 TB of data, overall, which will be easy to archive to user data folders.

You then move on to the *Plan* phase and document a plan of transferring MX (mail exchanger) records and mailboxes to the new IMAP (Internet Message Access Protocol)-based email system hosted by a top-tier provider. The design covers the needed storage for individual email storage on the SAN, and a 3-year Total Cost of Ownership (TCO), given the usage patterns. You present the plan (or maybe just a design summary) to management. They are less than pleased.

Why? You didn't ask WHY the migration needed to happen. You assumed you knew, and assumed it was based on cost. After all, maintaining your own email server is a cost to the company. You have done this sort of thing before and it was successful!

What you didn't realize is the reason to move "to the cloud" was to enable the workforce to be more mobile, requiring a secure mail and calendar application, so that any device could be used to access all of the information. Your *Plan* limited them to devices they already had, and only covered email. They wanted people in the sales organization to be untethered from the internal network completely! But—here's the key— they didn't say it in their project. They just said to "move email to the cloud." Yet, missing this bit of information about the business needs, caused everything to follow to fall completely in on itself.

Does this sound ridiculous? It happens all the time, and even seemingly mundane changes or upgrades can hinder future growth or make for a risk of re-work. The key is, to ASK and UNDERSTAND the business needs, in detail, from Day One, on every change.

We will talk soon about defining Success Criteria, but even before that point, you need to have a good idea of what the business's objectives are. This is often very uncomfortable, as you may feel you are upsetting leadership executives by asking clarifying questions when they feel they've been perfectly clear. It is best to use disarming phrases like "Let me be sure we are on the same page…", or, "Let me make sure I understand your needs…," because they work well in the United States. You may need to phrase this a different way in other regions which triggers the part of management's emotions that lets them know they are being heard, but you care enough to make sure your understanding matches their own.

The bottom line is, you need to fully understand the impacts of what is behind the drivers, and never assume people know what they are asking for. Consider examples where an executive hired a team to "deploy VDI"—Virtual Desktop Infrastructure—because he read about it in a magazine and thought it was a good idea. His staff was shocked at the request because they were already on virtual desktops, so they told the team to go home. Other times, where there is no real business need for a request, it is only useful for one department or team—sometimes merely a preferred tool no one else knows how to use. In worse cases, the tech is simply a toy or "cool gadget" that brings few others value. I cannot count the number of law offices I have seen this happen in!

The technology world moves fast, but accounting and budgets often do not. You may find midway through a process; an improvement or new release has been made that better fits an objective. But, if you do not have a firm grasp on what the business you are serving does, or how they do it, you may miss it.

For example, I was part of a project for a manufacturer to address concerns about the number of service desk calls coming from their sales staff. The complaint was the system was slow. The calls were consistently rejected as "cannot resolve" because the salespeople were connecting to cellular networks or unreliable Wi-Fi connections. They were instructed to work from home or the office. What the Citrix administrators did not fully understand is, the salespeople were taking orders live from the locations they found themselves, and not being able to accommodate their needs was actually costing the company revenue. The administrators my team was working with did not communicate this need to us properly, so we continued looking for ways to work within established standards to optimize the environment. When we interviewed management and found this was impacting revenue, we took a different approach.

Citrix had, at the time, just released an update to their ICA (Independent Computing Architecture) protocol (sometimes referred to as HDX), which would allow for a much more adaptable transmission of data by reducing the video frames and color depth. We investigated and found that the applications being accessed did not require much color depth and we instructed them to enable the new protocols.

The result was the sales staff were able to take orders and were much more excited to expand their reach in the field—all because we took the time to better understand the business justification of the technology. What could be accomplished well enough by other means (order entry at the office) seemed acceptable from an IT cost perspective, but the business actually needed more agility.

Here are a few business objectives you may encounter in project descriptions, all of which need further validation you will need to ask about:

Noted Objective	Example Questions to Ask
Reduce IT Spending	How much of a reduction is required? What areas or functionality are unacceptable to cut? Are there acceptable ways of shifting costs? What areas (for example, Human Capital, Licensing, Power, and Maintenance) are too expensive?
Create a DR Plan	What is the loss of time (outage) acceptable before revenue is impacted? Do we properly understand the difference between Disaster Recovery and Business Continuity? Are we talking about the same thing?
Build a VDI environment for Developers	What is the reason Developers need VDI? (Is it to reduce cost? Are they overseas? Is it for security purposes?)

Information Gathering Assessments

There are several types of projects that begin by gathering information about the company and systems in order to inform a *Plan* phase. Most of these are in terms of new technology or major upgrades, where a fair or large amount of technical data is needed to be analyzed. So, in the *Understand* phase, you may be called upon to start by using tools to gather this data. For example, if the project is a Workspace Transformation, you may want to use tools specifically designed to gather usage data from PCs. This information would contain the applications being used, how often, which users are logging in, and more.

From the data gathered during the *Understand* phase, you will typically determine the needs for the next phase to address. This means two primary skillsets are needed for these kinds of projects. First, you'll need an Engineer with the ability to work with teams that will be deploying collection agents, running scripts, and documenting findings. Fortunately, most Assessment tools provide their own documentation outputs which can be integrated with your own documentation. Second, you'll need an experienced Architect to review the findings, formulate the needed success criteria, align requirements against the stated business goals of the project, and organize the key findings into a presentation.

Examples of these types of Assessments are:

- **Transformation Projects**. The goal is to change the way work is performed.
- **Desktop OS Migrations.** The goal is to update platforms.

- **Virtualization Sizing Assessments**. The goal is to plan for the hardware needs for either physical to virtual migrations, resizing, or expanding the current environment.
- **Cloud Migrations**. The goal is to move workloads currently in the company-owned resources to a public cloud.

By no means is this an exhaustive list!

The "Day in the Life" Study

If you are in the *Understand* phase of a Transformation type of project, or even a Design review, your first goal should be gathering real-world information about how the users typically interact with technology to get their work done (or don't). The idea is to look for improvements which will help them meet the business objectives in your scope.

One of the most effective ways I have found to gather information, and also give greater confidence in this regard, is to conduct a "Day in the Life" study of your target users. Your goal with this study will be to determine the needs, implications, and challenges that users typically face. The goal during the *Plan* phase will be to meet this study with the business needs and objectives to determine which will be practical. For now, the task is to observe and report on what the users do in their daily work and how that impacts your project or initiative.

Just as the name implies, you will be observing users directly (perhaps assigning a specialist to conduct this task, at times). For a Day in the Life study, you will want to prepare ahead of time by getting a buy-in from your leadership or management. From there, you will typically need to find candidates for each target use case. This will typically be done by speaking with their direct leadership or managers to determine who would be best to work with. Be mindful that you may face some resistance if the users are very busy. If this happens, I would suggest having leadership talk to leadership. Your leadership has already bought into it, so let them work on the other leaders!

A "Day in the Life" may not always be an entire day study, but you want to be able to faithfully represent what their day would look like from an IT perspective. The exact information you are gathering will vary, but typically you'll want to examine or observe:

- **User start-of-work processes**. How are they authenticating? Are they using the system you expect or something else?
- **Applications.** What applications are they using? What do they keep open at a given time?
- **Performance.** What kind of performance are the users experiencing? Is it what you would expect?
- **Workarounds**. Workarounds occur when a user uses processes outside of the accepted instructions or expectations to get the job done. At times, these workarounds may violate company policy, so they are important to know. Ultimately, users want to work and if they find an easier way, they will take it. Be observant of these workarounds because they may be something you can *Plan* a solution for to help your users.
- **Challenges**. What are users struggling with? (Oddly enough, writing down things that aren't even in your scope can help you later. For example, if you wrote

down that they were complaining about the office being overly noisy, when someone requests a webcam for video meetings, you can exceed their need later by sending a headset as well because you know the built-in mic won't be a good fit.)

- **Use of Peripherals.** How often are they using attached peripherals for input, scanning, printing, etc.?
- **Mobility.** How mobile-capable are the users?

Note that observing everything is not always practical if you have a large distance between you and the users, but there are still parts of the study you could potentially adopt. Of course, if you can be there in person, you should try to do so. You will catch more subtle information and identify workarounds more effectively. Plus, it sets a great tone—that you care.

It goes without saying, you should be the best version of yourself you can be. Make sure you are well dressed, smell pleasant, and are in a good mood. Details matter!

Interview-Based (Infrastructure) Assessments

"What would you say… ya do here?"

—Bob Slydell, character in the movie *Office Space*

When a company needs to understand if they are correctly configured, have the correct procedures, or the right fits for their business's needs, information gathered will need to be both technical and observational.

Infrastructure Assessments (which I sometimes refer to as "Leading Practice Assessments") tend to be much more interview-intensive of the Operations staff than Information Gathering Assessments. They will require a project lead who has several years of experience and who is well versed in leading practices for several technology areas. The people conducting staff interviews will need to be good communicators but also good listeners who keep the conversation on track.

You should also be aware that as professional as you may be, the people you are interviewing will not always be. Often, when there is a known problem, you will encounter people that feel as though you are critical of their job, or they will wonder if they are being replaced or moved. This can cause a level of anxiety that you must be aware of. For this, I always think of a particular scene in a movie, released in the United States in the late 1990s, called *Office Space*. Two consultants, both named Bob (or "The Bobs," if you will), are tasked with finding inefficiencies in staff and processes. The staff, bored and unchallenged by their work, see this as "interviewing for their own job." Most of us wouldn't like to be seen as "The Bobs," but quite often that is exactly how we are being seen.

You will need to determine a personal style of disarming people. Back when the movie was more popular in the culture, I would often make it a point to introduce myself and my teammate as "The Bobs," but if you are sensing tension, that may not be best. Very often, in this kind of situation, I will remind each person I am interviewing that I am there to help them. I do not say I am "there to help the company" very much, on purpose. In the movie, *Office Space*, a large banner was displayed the day "The Bobs" arrived, asking, "Is this good for the company?" A very dull manager asked the staff to consider their day-to-day jobs as if they are doing things that are good for the company.

Let me tell you something crucial to your success in a technical assessment: the people you are interviewing do not care about what is good for the company. They are thinking of their families. Often, they are living paycheck to paycheck and are concerned the next one may not come. If you start by letting them know you are there to help them be effective in their job, they will fill in the blanks.

Save your "good for the company" language for chats with the executives and your Key Findings Presentation (the topic of another chapter).

In terms of understanding what your objectives are for an Infrastructure Assessment, you will need to be primarily concerned with "What" and "Why" types of questions. This means you will be asking questions or investigating configurations, determining the way in which things were configured, and why they were done that way. This will lead you towards the most important part of *Understand*, which is guiding the company towards positive *Change*.

Of course, the unfortunate thing that must be said here is a level of prerequisite experience is needed to make quality recommendations to a company. That being said, a team approach in Infrastructure Assessments often works very well. Junior members are either invited to take notes and learn, or the team lead allows them to lead parts of the conversation while supervised. I have seen success in using chat programs between the Assessment Team members to coordinate questions but be careful of this beyond this use case. The best thing to do in situations where there is a team is to first set expectations of the team. Discuss the day's agenda before the first meeting so each assessment team member knows their tasks. Splitting note-taking and leading conversations is effective because it allows conversations to continue while the person taking notes writes observations during previous and current conversations. If you are part of a larger consulting organization, as I was, you may not always know the person on your team going into an Assessment. If the Assessment is on-site, operating as a unified team is an absolute must to be successful. Just avoid the Michael Bolton jokes and you'll do fine.

Asking Leading Questions

Without descending into Chaos Theory, I will simply say, although certain standards exist in Information Technology, it is the differences and nuances of each company that keeps them competitive and unique in a marketplace. Because of this, you must recognize deviations from so-called best or leading practices are very often needed, to match other items of differentiation. Some deviations stem from practices which have been in place for decades and are slow to change. Others may come from inherited practices, such as from staff which moved from another company who did it that way.

I think of this story to put the topic of leading questions into perspective:

One holiday, a woman began to prepare a roast for her family. She cut off the ends of the roast and put them into the pan. Her husband, seeing this as odd, asked his wife, "Why do you cut off the ends like that?"

Puzzled by the question, she thought a moment, then replied, "I don't know. I always saw my mother do it this way, so I do it this way."

The couple decided to ask the woman's mother why the ends of the roast were cut off. Did it make it taste better? Did it make it moister? "I don't know," she said. "My mother always did it that way, but I don't know why."

Perplexed, the three decided to take the question further. They made the same inquiry to the grandmother. "Well, the only pan I had was too small to fit the whole roast, so I always cut the ends off to fit it in the pan."

A great many IT Departments will inherit ways of doing things and not know why. It is your task to investigate and determine the appropriate areas of risk.

To go about doing that, you must be aware of design decisions, even if a formal decision was not made.

You will need to be able to explain the elements of a current configuration and gaps that exist as well as be able to answer questions, to match the business's objectives of the project.

First, you must formulate what Leading Practices look like. This can be from your own experiences, but the best way to go about this is to engage the community as a whole. Thousands of blog articles, books, and videos exist for this exact purpose. It is your task to parse them and determine which practices lead the field in terms of effectiveness, lowered costs, and a lessened risk of outages.

Next, just as the husband and wife in the aforementioned story, you must be willing to ask questions. "Why?" may be the most irritating question to those with a three-year-old child or three-year-old SQL server, but it is at the core of every successful professional consultant.

Every deviation from what you know as a leading practice needs an explanation. This does not need to be confrontational, but unexplained deviations introduce risks. "Because we didn't know better" is a valid response, but you must be fully prepared to lead into solutions in the *Plan* phase by properly identifying all of the needs and challenges that surround the configuration in this phase!

As you can imagine, you will need to work towards perfecting a balance between consuming too much of a person's time and leading down the right paths with clarifying questions. I usually leave the ability to ask more open questions with people and, if questions come to mind that I haven't asked yet, I will often write them in the margin of my notebook or flag as a follow-up item in Microsoft OneNote with the "Question" tag. The key here is to get the person's information and follow up with them, if you are stuck or feel like there is more to be asked but you can't bring it to mind. Get their permission, of course, but this is the area where a peer review is important. By letting someone else into the experience, they will often think of questions you did not think of. Just because the interview is over doesn't make your need for the information any less important. Allow the process to work!

The Human Element

Another very common thing to keep in mind during Assessment activities and interviews is, if there are problems, there are a lot of human elements to be considered.

I cannot stress this enough: be compassionate but firm.

This is very often difficult for those without the perceived technical background to be an authority. If it were easy, everyone would do it. For those called to the IT Professional Services life, it is crucial to balance being "a friend" with being "an authority."

Many of the so-called experts I have spoken with do not possess the kind of confidence you would expect. It comes down to creating the right perception—a projection of confidence that you are responsible for getting the best information to those you are serving. This is true, even if it is outside of your current knowledge.

This is difficult to teach, unfortunately, but in my hard-learned experience, I have discovered some effective strategies to deal with these common challenges.

They may seem unrelated at times, but they all build towards being effective in giving folks the voices they need but using you as a filter to management.

Be an appropriate sounding board. Often, the people working on the systems are looking for a voice. Quite often, team members you will talk to are so surprised that someone is both knowledgeable and personable, they cannot help themselves. Here are some guidelines I use to keep the conversation appropriate:

> **Connect with others in a different setting.** Depending on the situation, I find that offering a post-work gathering where people can talk about what is on their mind is often valuable. Connecting with others can make the work more enjoyable.

It is not appropriate to allow off topic conversations to dominate a session, so here are some additional tips for preventing that:

Be mindful of the time-sink. Most interviewees will be cooperative, especially if they are looking for a distraction in their day. Many will be thrilled if an hour interview turns into four—they'll just blame you when asked about why they didn't get their other tasks done that day. Be mindful of useless conversation points and "rabbit trails," which are conversations that flow too naturally and end up off-topic. You can control this by setting an agenda at the beginning of each session and being very careful to split sections of the topic up by bio-breaks or to check emails, etc. At the beginning of each section, start by declaring the subject area covered. For example, "For the next 30 minutes, we are going to talk about your SQL Database infrastructure." End by declaring, "Let's take a 10-minute break to stretch (or have a bio-break, etc.) and we'll re-convene here to talk about Storage for an hour or so."

Disconnect the "who" from the "what." Very often, you will be in an assessment process because something is wrong and they are aware of your goal. Avoid a discussion of identifying who did something wrong. It is important to be aware that around eight of every ten of these kinds of discussions will involve a team member either pulling you aside or calling out another team member, a prior employee, or a vendor who "did it this way" and they "don't know why."

Keep in mind it is a potential liability for you to continue the discussion in a way that implicates anyone.

Engaging the conversation this way is unprofessional. More importantly, if you do, you are setting a dangerous precedent: when you leave, you will be blamed when the next interviews come up. This was another hard lesson for me to learn. In wanting to be "on their side" or get a staff member to open up to me, I used their language: "How did (X Company) set that up?" or "Why do you think (X Person) configured it that way? Did they leave you any documentation about that design decision?"

A few months later, I found out another consulting team was involved, and I was now being implicated in a similar fashion. In that case, it turned out a set of personalities were in play that were outwardly hostile to any outside help, but would play along to give consultants "just enough rope to hang themselves" or to ensure that decisions not made would be blamed, rather than good actions taken.

You will face toxic environments in Information Technology, so, first, keep to the methodology. Second, disconnect "who" did "what" things, and simply keep to the facts. Even if a team member mentions who did it, be mindful of it, but do not repeat it. Guide them away from talking about persons or companies and keep the conversation to what is there now.

Similarly, keep this in mind when making observations and recommendations as well! Try saying, "Active Directory Organizational Units should be structured to a more shallow depth to allow more efficient logons" instead of "The Organizational Units in Active Directory set up by (X Company) are inefficient and causing slow logons."

Be mindful of agendas. Pay attention to this if you are a consultant. You are not there as an employee but as someone present for a very brief time to accomplish a very specific goal. You are often perceived (as you can probably tell, from the other points I mentioned) as a potential mouthpiece for staff to push an agenda. You should not ignore this or dismiss it—input from those that are close to the situation is extremely valuable to determine courses of action and background. However, you must be aware of the "coloring" of the information they give you. Very often, management will listen to a consultant but not to their own staff. It is tragic, but it is an unfortunate reality of the business world. Many team members know this and will try to lead you towards conclusions to be their voice. In those cases, the most important thing to do is not make assumptions and ask the kinds of questions the people paying you to be there need you to ask. The bottom line is, even if the staff is correct, your job is to back that up by asking the right questions to justify the recommendation. Do not allow your desire to be liked or to keep the conversation flowing alter your judgment or cause you to take shortcuts.

The recommendations you make must be your professional recommendations, from your personal perspective. In other words, at times it may be appropriate to share what you would do if you were in their position – but be mindful that your words may be taken as the recommendation of a specific software vendor. Context of your recommendations are very important and putting them in writing will often make all the difference in perception.

I have some additional tips for those conducting interview-based Assessments.

Give people advance notice. Another way you can control the conversation topics and also be sure people are well prepared for you is to send a calendar invite that explicitly lists the timeframe, topic, and any information they can bring. This has the benefit of relaxing people a bit. Also, in someone's busy day consisting of several dozen appointments, yours stands out because you have prepared them. They will be much more likely to come to your session alert and ready.

Be aware of when to bring others in. It is okay to ask who may know the answer to a question. "I don't know" is a common response you'll hear during Assessments of this nature, and most of the time "I don't know" is said is because the person is trying to ask your permission to tell you who else should be involved.

Be the expert they think you are. The people you are interviewing are looking to you to guide both what is to be done technically as well as the conversation. Picture what it would look like if two people sat in a room waiting for the other to talk, neither sure if they were worthy to start the conversation. My best advice on this is to go in with a list of questions to start—items you know you will always need to document.

Chapter 3: Conducting Assessments

There are some important things to master when you are in the Assessment process.

The Six Keys to Quality Assessments

Observe and Report

Your first tasks in the *Understand* phase—which are critical to your success—are to observe and report. It is almost like you are a security guard who is not trained nor being paid to do anything but make observations; to write down what is happening and call the police, if needed. Similarly, you are not taking actions of any kind in the *Understand* phase. You will have a good number of findings, if you follow a good process, so the urge to act will be strong.

Remember, you are not making CHANGES yet (the next phase), nor are you allowing changes to be made until the full process is complete. This may seem counter-intuitive; it seems prudent to solve a problem as soon as you find it, but rarely does that well-intentioned action have benefit to a company. Quite often, when a proper procedure for change management has not been well instructed or is not followed (again, even if the intentions are good), disaster can occur because not all implications were found. Furthermore, trying to fix problems at this phase will very often distract you from finding more things that require your attention, thus lengthening the cycle and running a much higher risk of not finding key items.

You must remember that your job in *Understand* is to write down what you find and present it into a deliverable. Observe and report.

Avoid Assumptions

Second, avoid assumptions. Assumptions made will quite often destroy the validity of your findings. I could not tell you how many times I was reviewing a consultant's list of findings (after they spent hours on findings and recommendations), only to ask if they had documentation that the configuration was done incorrectly... and learned they did not. In investigating further, I'm amazed at how often consultants would assume that because they saw something one way three times, the fourth would also be that way. If it was not, that presents a large risk to the findings! I should not have to say this, but a professional does not cut corners. If you do, you are putting both your career and the project at risk.

Slow Down

Third, slow down. Take your time. I understand the desire to race to the finish and make the powers be very happy with your rapid progress, but the reality is, when you don't take the time required to discover the nuances of what is happening, you may risk those people deciding you are not worthy of the time they have invested in you. Worse, and I hate to break it to you, but doing something quickly is not always as desirable as people think. Stick to the plan.

Set Quality Expectations

Fourth, set quality expectations of staff involvement. The *Understand* period is typically the most interactive with more of the staff involved. The problem is, management may assume you are operating autonomously, unless you let them know otherwise. The process typically involves a lot more interviews, configuration reviews, and meetings than the rest of the phases of the methodology. So, it is very important to set scheduled expectations for those needing to help you understand focus areas. For example, let's say I'm assessing an Exchange environment. I will probably need to talk to teams who manage LAN, WAN, Active Directory, SQL, DNS, and storage, in addition to those in charge of the mail systems. While sometimes this is the same person, more often than not, this will be several different people. So, a good schedule and advance notice of the expectations to those team members goes a long way to being assured they are properly prepared and know the *why* of your Assessment.

Recognize Rapid Changes

Fifth, recognize that the world of Information Technology changes rapidly. New optimizations, security threats, and responses—along with new features and requests—all have effects on each other. Never assume that because a recommendation was valid once, it will be again just because only a little time has passed.

I cannot tell you the number of times, when in the midst of a project, a security change had been made which affected several of the previous recommendations in large ways. Or, in many cases, management will be upset that a recommendation you made a year ago is no longer considered a leading practice. It's not that you were wrong when you made it; it is that it is no longer considered "best" because a better or safer way has been found.

It is your job to properly educate leadership of these things in a way which maintains your marketplace authority.

Define Success Criteria

Sixth, do not forget to define Success Criteria! Especially during a Proof of Concept, project stakeholders need to agree with you on a key, but often overlooked question: What does success look like? It seems simple enough, but the project stakeholders might be put off at you asking the question. They may feel as though they have already defined it (see the second point above). However, taking a little time to ask that uncomfortable question goes a long way. Be patient and guide leadership in answering this question. Diagram, document, or "walk them through" the process to be sure everyone understands and are all "speaking the same language" about the goals. DO NOT skip it and assume everything will be okay. If the project is to change the way users are authenticating, a key success criteria may be to define the expected behavior of applications and what management expects to see. Will multiple challenges be acceptable or should there be a focus on single sign-on, where a user authenticates once per session, to all applications? It seems simple and perhaps obvious enough, but perhaps you can see if the intention is for users to be challenged for an authentication for security and you configure for single sign-on, there will be implications, including time wasted, so it is better to have that awkward conversation now than to run into problems later.

Strategic and Tactical Focuses

An often-misunderstood part of the *Understand* process is the difference between the overall timeline of a recommendation, the focus of it, and the importance to an organization a recommendation may fall under. Will the observation or recommendation need to be addressed to fix a problem, or will it be part of addressing an overall need? Will the observation or recommendation be something that will be isolated to one focus area, or require a good deal of effort in other focus areas in order to be addressed? Is what you are pointing out a weakness in training or overall staff processes?

These questions are not always cut and dry, and believe me, it takes a few months or sometimes years to fully grasp the impact of why we ask these questions this way, but in considering our audience, we must think outside of a single scope of maybe a group of engineers or Information Technology staff. We must make recommendations that address the company as a whole.

So, we group our observations into those that are strategic and tactical. Although these terms are essentially military in origin, they apply well to the business world, where there is a constant striving towards being better than you are today. One way I have heard this described at a high level is, strategy defines the right way of doing things. Tactics are the methods by which it is done correctly.

Strategic Focus

Strategy focuses on what is best for the company or organization over the longer term. Strategic observations (risks, needs, etc.) are those that should be addressed or considered as part of a longer strategy, such as actions which will positively impact the business's overall effectiveness or improve overall user experience. For example, they can sometimes be something considered as an overall update which will take other updates into consideration. Strategic items are those that tend to take months in scope, rather than days or weeks.

Another way to say a focus is strategic in nature is if it pauses to consider impacts on other systems, staffing, human capital needs, or will require a project specifically to address. Or, you could simply say that strategy is the "What" of the overall goal (the "Why").

Tactical Focus

Tactics focus on how to best accomplish a goal or overall strategy in the short term. Tactical observations are those that correct a configuration or act to protect data, personal safety, or other items that are targeted to a shorter timeframe, often within a few weeks or less of the observation being made.

Tactical focus tends to be for items that will either not affect other systems, will not require large amounts of effort to complete, or are covered as part of a previous project initiative.

Perhaps most important to consider is tactics feed into an overall strategy. If your observations do not fit within a larger strategic focus, you may want a peer review to check: Are you making a strategic observation and don't realize it?

Tactics are the "How" of an overall strategic goal, and the most common to need changes in recommendations. We live in a rapidly changing world. As I said before, never assume a recommendation you made last year is still valid!

Examples

Let me give some examples, as that is what typically helps me learn and I am sure many of you are the same way.

Observing 200 of 1000 users have passwords not set to expire is a *tactical observation* because it can be addressed quickly, without impacting other systems. Observing 900 of 1000 users have passwords not set to expire indicates a policy or other system requires a better design and is a *strategic observation*.

Recommending an upgrade of a domain member server because it is out of compliance is a bit harder to discern between strategic and tactical, but it tends to be tactical if it will have no other impacts. Upgrading a domain controller to get to a later Active Directory schema is a strategic recommendation because it will require a lot of coordination and effort. It will not be done quickly.

Recommending the testing of profile or persona management exclusion policies to not include temporary Internet files is usually a tactical adjustment. Recommending a company implement profile or persona management is strategic.

Identifying Risks

Risks are something I'll be covering frequently. Detail on how to write out risks will be given in the chapter called Risks and Recommendations (Chapter 10). Our goal with this section will be to work on determining risks.

A risk, in our context, is an unintended result of our changes that negatively impacts outcomes, safety, security, or business priorities. This can be something that is already present, something that is caused by changes that are being made, or something that may occur if changes are not made. Ultimately, you are attempting to identify anything that can be detrimental.

Let me make this clear: YOU CANNOT KNOW EVERY RISK. Your task is to identify those that you see. Everyone involved in a process should be mindful of risks.

There is a time and a place to discuss risks. That place is NOT while you are still gathering information. You need to get a complete idea of everything that is going on and ensure you've gathered all the information before pointing out risks to the company. As a consultant, I have made this mistake often. Once during an Assessment, I returned the next day to find the administrator I had been talking to "fixed the glitch" because I had mentioned it during our discussion the prior day. We later found out the administrator had not followed any proper change procedures and we were being blamed because "we told him to." So, learn from my mistakes, as I have.

Do not discuss the risk in detail in the moment or make recommendations now.

I will probably repeat myself with this at least ten times in this book to adequately get that point across.

I will now outline and discuss some of the basics I look for in identifying risks, especially during the *Understand* and *Maintain* phases.

How to Identify Risks

Study the way it should be. I read a story once about how Canadian Police were trained on how to spot counterfeit currency, not by studying fakes, but by studying the real thing for hours and noting the nuances. Having the right conversations and noting the way things should be (the leading practices) is a great way to point out things that are not right. In the same way, if you are focused on a technology area, do not neglect your education! Do not assume you will somehow encounter every scenario; you will not! Just as you are reading this book to learn more, do the same with the technology focus you have.

Borrow someone else's diploma. How do you know if something is a risk? Unfortunately, there is no real substitute for experience, but as my colleague, Jon Acuff, said, "It's okay to borrow someone else's diploma" (see https://it.justdothis.net/bl#2).

In other words, you don't have to be the expert in everything. Using your network of professionals, identifying a mentor, and studying under others are fantastic ways to garner others' expertise.

Listen to your instincts. Just because you can't immediately identify what is wrong doesn't mean your sense that something is wrong should be dismissed. Write out the question in your notebook. Ask someone about it. Search the Internet.

My guess is, more than half of the time, your instincts will lead you towards identifying a risk. This is the way our greatest computer, our mind, works. Don't be afraid to utilize it; just do so in the right context and not in front of those you are serving!

Recognize that your experience is valid. How would you do it? If someone is doing it differently than you would, is that valid? This mental challenge is hard for the engineer with less experience, but challenging notions is a fantastic way to grow, be it for identifying risks or otherwise.

The best consulting practices in the world have grown their list of leading practices, not from figuring it out on their own. They learn from challenging the way they would do things when they see customers doing it a different way. It is not making assumptions that they always know best, but relying on their experience to formulate a leading strategy.

Watch for updates. I have touched on this a lot so far, but leading practices change because we are working in a world of rapid improvement. Think about it. How fast has train technology evolved in the past 100 years? (Spoiler: not a whole lot.) Yet, risk assessment is a key part of operating a railroad. They need to know what has changed, or might change, even if it seems not much has. How much has changed with technology? A lot! In fact, I recommend at least five to ten percent of your week should be spent reading updates, forum chats, and other current news about your technology focus because updates often involve improvements or patching security, which is one of the most important risk areas you can identify. Frankly, often the risk of not staying up to date is losing a competitive edge which could reduce costs in the company. I think of how many times I have been in front of customers who have employees that literally interact with each other only two to three hours per week. Their workloads can probably be done anywhere, so if the company is spending $2000 per employee, per year, to have them in an office, and they need to expand to a new space to house 1000 employees, they have a risk they don't even know about. The risk is, they could be saving hundreds of thousands of dollars by allowing their employees to work from home, but they didn't know it was possible or feasible.

If you don't know about improvements that have been made, neither will they!

Avoid timidity. "If you see something, say something" is now a common saying because of the "If You See Something, Say Something™" national campaign set forth by the U.S. Department of Homeland Security (see https://it.justdothis.net/bl#3). It is believed the terrible events of September 11th, 2001 might have been lessened or perhaps even prevented if people had noted to authorities that something was wrong and let them determine a course of action. My thoughts go to the movie called *The Boondock Saints*, where a priest told his congregants that evil will prevail if good men do nothing.

It is important to find a balance in the moment and know when to point out risks and how. The key here is to get used to pointing them out and not assuming someone else knows or has done so. This will be something that is developed over time. For now, my encouragement to you is if you see something that may cause a problem now or down the road, say something!

Write it down. Pointing out risks is something far too important to rely on memory alone. Writing down what you see allows for collaboration later in the process. If you are in the Understand phase, a lot of data is going to be coming at you at once, and it is important to keep it flowing without slowdowns. Keeping a notebook and making quick notes is a great way to follow up with things later and not slow down the process in front of you or distract others. It also sets your mindset properly to look for similar risks in your process as you go along.

Be aware of opinion versus risk. Frequently, seasoned engineers will give opinions of how they would rather see things as "risks." Hopefully, you are familiar with the fable of the boy who cried wolf. If every small thing or variance from your way of doing things is categorized as a risk, it will dilute the impact of your work. Ask yourself, *Can I quantify this variance as a risk, or is it just how I would do things?* In the process of writing it out, if you can't identify why a variance you have identified would be beneficial to address, this is probably something best saved for another section of the document, or left off completely. We always want risks to be quantifiable.

Organizing Risks

Now that you have some risks identified, it is time to dive deeper into making them actionable. First, however, we need to expand how we will describe them.

Categorize. As we progress and write the risks out, we want to sort them. Identifying a small list of categories is best for this. Overall, you'll want the list to be limited and relevant. Don't make up a different category for each new item. For example, you may have your risk categories defined as:

- Security
- Usability
- Functionality
- Data Integrity
- Business Continuity
- Quality Assurance
- Operational
- Financial
- Return on Investment

Be mindful of impact. So, you've identified a risk. How widespread is the problem? The way I typically recommend looking at this is:

- How will this risk impact users, data security, or other factors? You will need to determine a level of impact and the implications of the impact. For example, if you find that two of four web servers in a round-robin load balancer are giving inconsistencies in the home page, you would be able to safely say that users are impacted about 50% of the time. Other risks could impact everyone in some minor way, but the impact needs to be noted. The majority of risks I have identified only impact a portion of the user population or data involved. We will score the actual impact at a later time. For now, it is important to get an idea of how widespread the issues are.

- Next, identify if there is a quantifiable revenue or cost impact to list. Of course, be careful with these kinds of estimates and always follow up with someone to check your work.

- In some cases, impacts may be external. For example, risks of a medical system malfunctioning may pose a care or life risk, which may not impact many people but the amount of impact per person is extremely high. We need to note these kinds of risks as well, and not assume the risk is known.

Prioritize. Another key in looking at risks is knowing which ones are more important to address and which will need to be addressed as part of a larger process. For each risk, you will want to identify if it seems like a strategic or tactical risk and write that down. This is something that is going to be important as we follow the Methodology in the iterative phases. When we score and sort these risks later, we want to have enough data to be able to assign higher priority to higher level of risk. I typically start with Low, Medium, and High risks and find a balance later if they need to be scored by number. An important morsel of knowledge here is it is possible to have a Low Risk Priority on a High Impact problem!

Let us use the web server example, again. Half of the time, users are being presented with inconsistencies in the home page. We have to ask: What is the inconsistency? If it doesn't impact usability or data security, is it really something the company should ignore a potential data breach to address first? My guess is not if it is something as simple as, say, a font placement on a picture caption, or something similar.

Your task is to dive into each risk and determine how you would identify the level of risk by better UNDERSTANDING the problem. Think of what you would want to be said to you if you were the executive that asked for an assessment.

Do not let setting the priority be "a hill to die on." Do not become emotionally attached to priorities you set. In light of potentially hundreds of risks a company faces every day, what seems important or unimportant to you may be re-categorized. In fact, you'll find we may reassign priorities even within the project. Right now, your task is to gather information for assembly later!

Identifying Recommendations

Just as with pointing out risks, in almost all cases, you must refrain from making recommendations early! This is a challenge if you are in hurry, have a lot to do, or just simply feel a need for the staff you are working with to like or appreciate you.

Recognize your instinct to point out helpful things is good!

I am willing to wager your nature is to help make things better than you found them, or you would not be reading this book. But it is important to be prudent, to truly be helpful. You must first recognize that nature and the implications of allowing your personality and knowledge to not be disciplined if you are going to be successful.

It is also important to note, you may not be the person who ends up making the final recommendations. In many cases, you will be working with a Senior/Principal or other Architects that will take the risks you have identified to identify solutions. This is something I have encountered a few times in my work with Citrix Consulting, when I was working with a Junior Consultant who was eager to please but confident. They will sometimes make a recommendation that is not correct to a customer in the moment. It is an awkward situation to have to correct another consultant in front of the customer. It damages confidence with the customer and with the Junior Consultant as well.

In bad cases, I have seen customers request different consultants be assigned because of this behavior. If you either are that senior person, or want to become that person, I have a few recommendations on how to go about that at https://it.justdothis.net/bl#4.

Now that you have some risks identified, what should be done about them?

Just as we did with risks, we have some work to do in making recommendations.

Recommendations must be actionable. Would you take your car to a mechanic who said you needed a new transmission but only stated the problem, not the solution? No! You would want a mechanic who gave you steps to take to solve the problem, even if it meant going to another recommended mechanic! In the same manner, you should be focused on making recommendations your audience is either capable of doing or pointing out who can help them with the problem (especially if that is you!). The most important aspect of this is that you are helping them to take action, not just be aware of a problem.

Think about the scenario. You should be able to envision present and future implications when you are making recommendations. It takes some practice and it definitely takes patience. Slow down and think about what you have learned and how to properly convey it. Writing down a paragraph for each risk—complete with links and references—is helpful. We'll work on shortening it in the chapter on Documentation (Chapter 9). For now, write it all down. Make it a story of what could happen if you are not listened to. What will happen in a year or two from now if they enact your recommendation but don't fully understand it? What happens if they don't do it at all?

Use a formula. This may be one of the most important paragraphs I've written! Using a formula is the first step in speaking the language of the executive, project manager, and engineer alike. Therefore, it is crucial, both for clear thinking and in communicating the issues well.

We will cover how to formulate this into declarative statements a little later, but the best way to begin with recommendations is to think in terms of the basics:

- How should the problem be solved?
- Who will solve the problem?
- When should/must the problem be solved?
- How much time/materials/cost will be involved? (optional)

Borrow experience. Just as in the "borrowing someone else's diploma" concept I mentioned for identifying risks, the same can (and should) be said for making recommendations. Very rarely is it appropriate for you to simply assume a recommendation is at the best it can be based on your opinion alone.

If you collaborated with others on nothing else in the entire project, now would be the best time to do so, especially if others have taken my recommendation of purposefully spending five to ten percent of their week learning about updates to their technology areas.

If you get in the habit of asking others to assist you, you will end up with a massive pool of current expertise!

Make a recommendation for every risk. Depending on the practice, you may be instructed that every risk must have an accompanying recommendation. While I think that is a good thing, I firmly believe you should not ever fail to point out a risk just because you do not know how to fix it! This point in the process is not appropriate to dismiss a risk you have quantified, even if the only action you can come up with is "Company will need to follow up with the Vendor to determine a course of action to address Risk before time."

Identifying Success Criteria

In the beginning of this section we discussed Success Criteria and why it is important, but here we will add another reason: Success Criteria will ultimately affect your testing plan.

During the *Plan* and *Change* phases, this may undergo modification, but in Zig Ziglar's typical "aim at nothing and you will hit it" way of saying things, if you do not define for everyone involved the Success Criteria, the Testing Criteria, and a Testing Plan, then determining when a project is actually complete will be difficult.

I have been part of larger projects that spanned nearly a year to complete. Imagine the surprise of the consulting team when they found that the person who they had discussed Success Criteria with verbally (but had not written down) was no longer with the company!

The new person had a different idea of what the project should be, but it was not communicated. So, with a fresh new environment in hand, it was handed over to the person's team to "test," and—I'm sure you know where this is going—the project was a complete failure. The team was required to make huge adjustments, for free. The consulting company ended up losing money on the project in the end. All told, it was a tragedy that could have easily been avoided with just a few words written in a document.

If the Success Criteria had been written down, and if an agreed-upon testing plan been in place, the team could have conducted the specific testing prescribed before handing the environment over. If the opinion of what "success" looked like to the new executive was determined at that point to be different, a scope change could have been made.

What does Success Criteria look like?

Interestingly for us, it ties in with a testing plan. The Success Criteria must be clearly understood.

"It works" is not a success criteria.

"Launching Internet Explorer successfully from an Apple iPad" is closer.

"Sales Staff are able to view the Intranet site from Internet Explorer using an Apple iPad from customer locations" is more precise and leads to our next step of defining Testing Criteria.

Testing Criteria expands this to give more specifics as to how the testing will be validated as successful.

In our example, we may want to get specific to say a certain group must have validated that they can access and manipulate their departmental files that stay on the private Intranet using a 4G connection on an iPad without downloading files.

For each Success Criteria, define how it will be validated or tested with Testing Criteria. We will report these along with a draft Testing Plan at the end of the *Understand* phase and update it in the *Plan* phase.

Chapter 4: Taking Notes

"Strike while the iron is hot."

—Sir Walter Scott

We have talked a little about notes. This section is going to give you more pointers and tricks, although you'll need to determine what works best for you.

The key to any note-taking is to make sure it will be valuable to you later. Well-written notes that others find valuable are great, but are not crucial in practice. Ultimately, the goal is to have ways for you to organize information to move the project forward.

Sometimes, notes are nothing more than screen clippings or saved sections of web pages. Other times, they are what comes out of discovery meetings which last for an hour or more. They may even be sales information. However, without an organization style that works for YOU, they are devoid of value.

How I Take Notes

As I covered in the Layers lesson in the Introduction, for nearly every type of IT project I have ever done, I have been able to identify layers. I have gone so far as to create OneNote templates for certain project types with a structure that looks something like this:

Tabs (Sections)

- General
- Understand
- Plan
- Change
- Maintain

This will allow you to use the same place for note-taking, following the entire lifecycle, in the same overall note file. Further, within OneNote, you can link to previous notes easily. At times, you may have multiple projects at once or projects that span months at a time, so the efforts toward organization early in the process yields time-saving benefits in the end.

In the General Tab, I first place notes about the business I need to know. I will use this for the Executive Summary and look back at it frequently for reference. Here, I also place notes from meetings, logistics, and other notes that don't have a place elsewhere. This is an appropriate place to keep follow-up items.

Layers (Pages)

At the beginning of each Phase's tab, I will typically have an overall schedule and other notes about the sessions. Put a note for each Layer under those. I will quite often include some "cheater notes" in this primary note, where I'll reference the topics that are going to be covered. If it is an interactive project, I will note how long each session is scheduled for and list typical questions to ask or items that must be covered.

Each of these Layer notes will have sub-notes under them for the individual notes. For example, my structure may have looked something like this within a tab in 2016:

Business Layer
User Layer
Access Layer
Resources Layer
     Applications
     XenDesktop Images
     XenApp Images
Control Layer
     Active Directory and Personalization
     Databases
     XenApp Farms
     XenDesktop Sites
     Provisioning Services
Hardware Layer
     Storage
     Network
Operations Layer
     Support Staff
     Business Continuity
     Security

Obviously, this is an evolving template.

You should be thinking in terms of using these notes in your delivery documentation. This does not mean you should be literally typing your documentation, but type enough information to trigger a memory and make filling in the details easier.

Here's a handy tip: if you are typing a lot of your notes, you may not want to type them twice. You can match the text styles for "Normal" and "Headers" to match the document font in Word. This will allow very easy pasting into the document, then elaborating fully in your documentation.

Starting half-finished will make you more efficient! More efficient people make more money and spend more time at home!

If I am using a paper notebook, I will very often take pictures with my phone and send the photos to a OneNote page. Regardless of if you do this or not, another important feature of OneNote is the ability to create tags for items, as well as customize those tags. I will typically make a custom set of tags with custom keyboard shortcuts and color coding. In fact, having a universal color coding system for each layer keeps things interesting and easy to reference!

Once you have key findings noted on each page, you can then create a summary page with either the entire section or with the entire notebook, if you wish. This can be very useful in preparing the Key Findings Presentation or just to have a central point of reference with all the things that should be done.

On occasion, it may be appropriate to skip the notes and start working directly on the deliverables. There are a few projects that lend themselves well to this—typically Health Checks and Assessments—when working as a team, though sometimes Design projects are appropriate. I say "on occasion" because I will more often refer back to a OneNote page than look back at a deliverable. That being said, I cannot dismiss the power of online collaboration of documents that is possible now. Just as your OneNote can be shared in real-time, your team can be working on a deliverable in real time. It is best to create templates that can be uploaded to the cloud for collaboration purposes.

One word of caution here: when I say "template," I mean just that. I have seen far too often Junior Consultants use previous work as a "template," thinking they would simply replace the text as they went along. That sometimes works, but in a collaborative, note-taking way of doing documentation, it can be a nightmare because you will not know for sure what others have done and what you have done. So, either make your template blank for each section, or have "example text" highlighted in such a way it is obvious it needs to be removed.

I also say "on occasion" for another very good reason: Your priority in taking notes is staying present in your meetings and information gathering process. Be aware of mental context shifting. When you have a document template in front of you, you are typically thinking of completing a document, which is a task typically left to a context of focus—perhaps using a laptop in a hotel room with headphones on cranking speed metal, crushing cans of Red Bull on your forehead. Or is that just me? Okay, fine. Regardless of your style, it is important to be aware of this context shift. If you are thinking about completing a document and you are taking notes in the same document, you may find the customer nodding off as you type, backspace, type, and try to get it right.

If you are working on your own in a project, I do NOT recommend taking your notes in the same deliverable document. Trying to lead a conversation is difficult enough for most IT Consultants to master. There is a difference in brain activity between these kinds of activities, which makes it both mentally difficult and physically tiring do to so. This is particularly true for the context shifting challenge because you will often hear something that belongs in another section. It is FAR faster to get into the other notes using tabs than to try to find where you should be in your deliverable.

Regardless of this, if you are working alone, the conversation is going to pause while you find where to record your note. Even the most skilled in carrying on conversation and writing will have this challenge—myself included. So, it is best to work on rapid context switching rather than expecting your client will be patient while you fumble with your notes and fade off into uncomfortable silences you don't realize are there. If you don't believe me, record your meetings and play them back. You'll hear it. I say this because that is exactly what I did, and how I realized the problem.

Document Early

"What is this? Is this someone else's document, D.J.?"

It was one of the single most embarrassing moments of my early career.

I had a stack of assessment documents, about four weeks deep, from five different clients, and I was way behind. While I could place blame on scheduling, the fact is, I was lazy for the first two weeks and got behind.

One week, I was desperately trying to get caught up. It was the week I became most thankful for the QA (Quality Assurance) process at Citrix Consulting.

One of the Enterprise Architects was reviewing the document I'd just sent for their review, prior to sending it to the client—a standard procedure, thank God.

The problem was the document had entire sections of issues I recalled over the week-long assessment. I had typed them up with masterfully-crafted recommendations that would save the client thousands of dollars each year.

The problem was the recommendations were from Client #3, not Client #2. The two projects were very similar, but I had relied on my memory in my documentation blitz. (I would later refer to the entire period of time as the "Summer of Hell," from what was required in terms of time, travel, an embarrassing re-write, calling the client to get clarification, and making sure I got it right.)

Regardless of my mental fatigue, that whole situation was preventable.

For the record, once I learned what you are about to learn, I was able to conduct an even more intense and profitable series of projects (seven in one month)—not that I ever recommend doing that, but sometimes you have to "stack" and you need to learn strategies to get it done.

Some of the best advice I can give in the documentation process is one that has come with hard lessons over the years and I would encourage you to take seriously: strike while the iron is hot. In this case, it means that while the discussions and information you have gathered are in front of you, do not wait to begin writing things down. If you have followed my advice so far, what you will end up with is either a document with raw notes or a OneNote page with notes—or sometimes both! But the thing about notes is they are only as good as your memory retention.

If you are a traveling consultant, as I was, you may quickly find that one IT environment runs into another in your mind, and your memory may not be as reliable as you think it is. If you wait until a week or two later to write up your documentation, you may run the risk of making a recommendation that is simply not valid at all—as I did!

To combat this, I learned that taking the time to work on documentation each day is important. If you had three sessions today about the User Layer, and your next discussion is in an hour about the Resource Layer, take the time to start making your notes into documents now. This is why a format for the document is agreed upon early in the process—it frees you up to begin working on the document in advance.

If you are working with a teammate, you should already have divided up sections to work on or agreed upon how you will contribute. Even a 15-minute discussion after an Assessment session can work wonders. Ask the other person(s) what they feel are the key findings of the session. Agree upon who will write them into the document. If you are working alone, ask yourself the same question and mark the key findings. You will need them later for presenting them to management.

Don't forget about context shifting. If you want to be truly effective at documentation, my advice is to schedule two hours each day for that purpose, if you can. So, if you have a 10-hour day, use six to seven for meetings and information gathering, one for summarizing key findings and notes, and the final two for being "heads-down" in the documentation process. Be serious about this! A mistake I have seen many Junior Consultants make—and one I have made quite often—is billing a 40-hour week with four days on-site, but they squander the two hours at the end of the day going out to eat, over-indulging in the local "scenery," or sometimes just catching up on whatever *Game of Thrones* was all about. I will tell you, the consultants who did that often found they actually worked more like 50-60 hours for a project because they had to back-track the following week. If they had just simply taken the time to stay on-site a little longer and worked in advance on their documentation and Key Findings Presentation, they would have been much less stressed when it came time to do the presentation three hours before the flight left. They would have spent less time asking me if I could lead the conversations *and* take notes for them so they could sit in the midst of a conversation with a client, while working on another client's documentation. Don't be that person!

Now, I don't want you to think I'm saying you need to work all the time. If you are traveling, for crying out loud, go enjoy where you find yourself. Just remember, without good discipline, you may find yourself in Miami, in your hotel room, working on the project you had last week in Delaware. Your time bar-hopping in Delaware was fun, but the bill always comes due, so strike while the iron is hot.

Documentation itself is important enough that it will have its own chapter. For now, your goal should be to get down as much information as you can, each day.

Do not worry about editing or correcting grammar, etc. For that detail and exactly what should be in each document, see the Documentation chapter (Chapter 9). I will also have specific information to include in each Layer's documentation, as appropriate.

Chapter 5: Presenting Your Findings

"Practice makes perfect." This is a phrase that was first seen in the late 1500s and is an adaptation of the expression "use makes mastery."

The Key Findings Presentation

An important part of completing the *Understand* phase is showing what you have found and validating with those you have been coordinating with thus far that you have the key information down properly.

This can take many forms, but over the years I have found it best to schedule a thirty-minute to a one-hour meeting to both show and discuss your findings. For these, I strongly recommend having a presentation displayed on a central screen. Sometimes, this will take the form of either a shared, in-person meeting or a remote meeting. In-person meetings are always best.

I typically recommend using PowerPoint for this presentation rather than simply running the stakeholders through an incomplete document that is hard to read on a screen. There are a lot of reasons for this, not the least of which is the client doesn't typically want to see what you are working on—they want to see what you have completed. If you show them portions of your document, you will get asked, nearly every time, "Why isn't this done yet?" Simply put, they will be less confident in you. However, you need to give them something. This is why the Key Findings Presentation works so well. If they ask for something, you can send them a copy of the slide deck. It is pretty much worthless without your words (and if you are smart, you'll record your presentation and send them that), but it puts management at ease to have something to show after a hard week of meetings that took their staff away from their other duties.

I have often been asked another question by my students and coaching clients that I thought I'd cover it as well: If the client asks for a draft document, what do you say?

The response is a magic word that has transformed my career.

Are you ready for it?

"No."

As in "No, we agreed our deliverables for this project were a Key Findings Presentation and formal documentation. A draft is not a formal document and it is against our policy to release draft documents that have not gone through our QA process."

That said, if your policy is not there, it should be! As well, you should neither be so unconfident that you are willing to bend to pressure and show what you are working on, nor should you be so prideful that you do not have a peer review process.

A presentation is often the "make or break" component for your project data, status updates, or closeout meetings. I have been part of companies that rush this process or cut corners. The problem here is, if the client senses this, they will not take you seriously. Do quality and thoughtful work. Allocate time ahead to give a high-quality presentation.

Your goal in giving a presentation should be to have what I call "participatory authority." What this means is, while people are participating with you in an overall discussion, you are using visual and other cues to control the flow and timing of the discussion. Participatory authority is a learned balance between dominating a conversation or process and being a timid listener. (I always think of the frustrated administrator portrayed in the 1990s by *Saturday Night Live* who says, "Okay, MOVE!")

This is also relevant during interview sessions because you need to be a good listener, when the time is right.

You've already done a lot of listening and data gathering. Now, it is time for you to present it back and let your audience participate in confirming what you have learned. That said, learning the skills to lead them to your conclusions without "hitting them over the head" goes a long way. Let them participate in your authority!

Key Findings Content

What should be put into a Key Findings Presentation?

As with most things for an executive audience, include as little as possible while still conveying the importance of what you are saying. Keep it tight! Doing so will increase your participatory authority.

Here are the nine components I typically include in a Key Findings Presentation, which align with the peers I have interviewed and gotten the most complements from management:

1. Agenda. This is vital to have, especially if you have C-level executives in the audience.

- Get the audience's "permission" to agree to the agenda using phrases such as: "I've put together a presentation to make the most efficient use of our time today. If it is okay with you, I'd like to hold questions to the end, just in case anyone has to leave." This is a disarming way of basically asking everyone, including the eager executive, to keep quiet because it isn't you that will be inconvenienced, it is the rest of the audience. An agenda seems obvious to you and maybe pointless to the audience, but knowing where you are going is crucial for them to participate in your authority.
- Encourage them to take notes. In fact, in the invitation for the meeting, if you can remind people to bring something with which to take notes, it goes a long way to helping.
- Handouts are sometimes effective, but don't always add value. They can take a lot of time and QA effort you may not have available.

2. Project Summary. Remind the audience why you are there. Use a slide that contains the business's objectives. In many cases, you will need to introduce yourself and the team involved. Highlight the client's team if you are there as a consultant. Don't assume your audience knows who they (or you) are.

3. Project Status. List what was done so far at a high level and what is still pending. (We will get into what needs to be done in more detail, at the end of the presentation, in the "Next Steps" section.) Use caution here. At times, you will be confronted with a "get to the point" executive who just wants you to skip to the end. Remind them that you have the time scheduled and you have a specific agenda to cover and want their cooperation in sticking to it. Keep in mind, many executives in attendance may have no clue about your schedule. Many times, I have had to explain the project entirely to executives that were invited to a meeting because someone over or below them just thought they should be there but did not give them context.

4. Environment Overview. Give a high-level view of what you have observed in this process. This should not be overly involved. Often, a diagram will do for a slide or two here. Keep your audience in mind for this.

- Executives will be frustrated by an overly complicated drawing.
- Engineers will ask a lot of questions.
- Using phrases like "conceptual" for diagrams is disarming for both. Remind the audience there will be far more detailed diagrams in the final documentation. I often joke that I wanted them to be able to actually read the text on the screen, which

brings up the next point: this is an OVERVIEW, so less is absolutely more.

- Condense, condense, and then condense again. The overview should draw attention to the findings.

5. Findings. Findings should be brief descriptions of what you have observed and what items need attention. Note that you should not be including recommendations at this point. There are several schools of thought on this, but here are my favorites:

- Overview—This style would be an overall health ranking or general findings, consisting usually only a few slides (at most), to guide conversation.
- Layers—List the top five most important findings for each layer—good or bad. This typically means a slide for each layer.
- Red/Amber/Green—Using a well-known visual queue such as a traffic light, use three slides: what areas have low risk (green), what areas have moderate risk (amber/yellow), and what areas should be addressed before continuing the project or have high risk (red). Try to keep these recommendations to high-level themes, if possible.
- Top 10 or Dirty Dozen Risks—Depending on how well you know the audience, you may want to consider a "risks ranking," either in a countdown that keeps them interested, or as a list that builds their anticipation. This is one of my favorites because it can showcase your value in the executive's mind, who mostly just wants to know what are the most important things to do and how much they will cost.

6. Data. In many assessments, you will have some sort of data to display to accompany the findings—perhaps a graph of support tickets or a grouping of tables and graphs describing the summary of your Workspace Transformation Assessment data.

- Find what data makes an impact towards your action steps and focus on those. Using our example, if you observe that logon times are slow, contrast them against industry standards to help your audience comprehend that your next step action of GPO tuning has merit.
- To establish participatory authority, consider showing this information on one slide and then on the next slide show a breakdown of logon steps that draw their eyes towards the issue (slow GPO application, in our example). Show it, but don't say it until the next steps.
- Be aware of the length when building out these slides and also be very aware of size of the data displayed. If you can't read it on your screen from a few paces back, your audience may not be able to see it well, even on a larger screen.

7. Next Steps. This is perhaps the part of the presentation that most stakeholders most want to hear.

- What are we doing, or what should we be doing about what we've found? This is where you will want to outline any defined next phases in the project.
- You will also want to match the style of the findings to these recommendations and next steps. For example, if you used either a Layers or a Red/Amber/Green style, you may want to consider the next steps be defined in a tactical slide and strategic slide. Be sure all the risks you mentioned are covered. If you used a Risk style list, use the

same numbering and outline how each one will be solved.

8. Summary. Wrap it up with any additional information. Verbally, you'll want to mention any deliverables that are pending. I do not recommend putting deliverable dates in a presentation slide unless you are absolutely required to.

9. Questions. Finally, include a slide declaring a thank you and asking for any questions to finish your time. Hopefully, if people were taking notes, this will be effective. That said, in my experience, if you have instructed people to wait until the end, you will get less questions and a lot less time spent saying things like "we will cover that in the next section."

Presentation Pointers

Giving presentations is often the most uncomfortable thing for someone who has an Engineering background to do. However, I will say that a technical person that can communicate well is destined for greatness. So, how does one get there?

Participatory authority. Remember, you are there to be the authority in the room. If other people knew what you were presenting, they'd be doing it already! So, practice being the authority who cares and listens. Be confident in what you are saying. If you are not confident in what you are saying, it is probably better to not say it until you are. This can often be done by allowing those to whom you are presenting to lead towards the conclusion and supporting them in it, but I would be cautious about using that trick too often in a presentation. It is far better to lead them towards a conclusion you have already made and intend for them to get to. That said, be very aware of the temptation to try too hard to convince someone that you are valuable. Be aware of the needs for acceptance and respect. Practice leading others to the conclusions you have made. You would be shocked how empowering others will boost your career and have management demanding to have you involved in their next big project! So, make it about them, not you.

Be brief. When people get nervous, especially those who are not used to giving presentations, they will often say either too little or far too much. In the same way you write, be concise and to the point. Remind the listener that you will have a lot more detail for them to read in the document. The last thing you want to see is people who you are trying to communicate with checking their devices or watches while you are speaking. Those actions are indicators you have lost their attention, typically because you have become overly verbose. Be brief in the slides, as well.

Avoid "technobabble." Technobabble is a phenomenon which occurs when someone is describing things that may or may not make sense to them in order to sound impressive or silence others. It involves using buzzwords, technical terms, and even tech slang—and it sometimes even occurs accidentally. You may do it and not even realize it. The last thing someone listening to your presentation wants to feel is inferior to you, which is what technobabble ultimately makes people feel, so it is crucial to think about what you are going to say before you say it. It is often helpful to make notes in the PowerPoint slide notes or on note cards. Think to yourself, *how would I describe this to my spouse or a* *family member?* The key is to remember your audience. In this session, your audience is typically not the technical team, even if they are present.

Practice. Practicing in front of a mirror only went so far for me. For a time, I tried rehearsing presentations in my hotel room, which was helpful, but what I began doing that was most helpful was the same thing that was helpful to me in college radio: recording my presentations as they were being given. First, I did this using only audio. Then, I did it using video. What I was listening for were the times I had awkward pauses and used gap-filling words like "umm," "uh," "so," and "you know." Very rarely did they show up during practice, so I kept working at it until I was aware of when those kind of things would be said and quite literally bit down to keep myself from saying them.

Use the technology. In most cases, you will find yourself hooking a laptop to a projector screen. A key feature to PowerPoint is the ability to use a second screen to have the presentation, while the primary screen has your notes, a view to the current slide, what is next, and the best feature of all: a timer. I began traveling with a second (USB-powered) monitor to practice with and ensure I had a second screen, even when presenting remotely.

Turn everything else off. Picture the scene: you are in the middle of a presentation and a chat window comes to the big screen. "Can you believe these idiots?" the chat says, visibly, from someone actually in the room. Aside from being a potential "resume-generating opportunity," it has some elements you should always be aware of: you cannot control what people will send to you, but you can control when you see it. So, close every program that isn't part of your presentation. Don't be the person in our example! The same goes for phones and other devices. Turn them on "silent," put them into the "do not disturb" mode, or completely turn them off (power them down). This serves to make your audience the most valuable people in the room to you, and you will be amazed at how much better your presentation is when you are not distracted by what may pop up. Encourage your audience to do the same.

Don't read the text on the screen. A major pet peeve that distracts many people is when you are reading what is on the screen. Most executives think to themselves, *I can read this on my own. They should have just sent a memo.* The next thing you know, your audience is distracted and not really listening to you at all. I have literally seen the president of a large corporation fall asleep during a presentation because of this one thing, so don't do it. Use your words to add color to what they see on the screen. If you engage multiple senses at once, you will be much more effective. I have even gone so far as to have no more than six words on the screen at any given time to ensure people are focused. In some cases, you may want to use the "blackout" feature, if you want your audience to look away from the screen and to you, which brings me to the next point...

Make eye contact. Unfortunately, this is another tough one for many engineers and administrators who have become used to working in cubicles, remotely, or generally not face-to-face with people. Remember, your goal here is to communicate with real people, so be real. Look them in the eye. Practice making sure everyone in the room has met your gaze at least once. This can be awkward at first. However, eventually, you will note that your participation goes up and people take you much more seriously.

Thank attendees. Everyone in the room is likely being paid to be there but it is still important to thank the attendees for their time. I have heard it said that "time is the only real currency," and I agree. So, thank those who have come for being there and use language like, "Thank you for giving me your time and attention today." You will be amazed at how simple changes such as that will improve everyone's experience.

Now that you're done with the Assessment, you can confidently say you are ready to move past the *Understand* phase. You have your actionable intelligence, next steps, success criteria, and everything else to proceed on to the next phase: *PLAN*.

Before you move on, check that you have done the following:

- ✓ **Delivered all the deliverables you were supposed to**. For example, if you are to write documents, make sure it is done, reviewed, and finalized.
- ✓ **Defined the Testing Criteria**. How will Success Criteria be validated? Who will be involved?
- ✓ **Communicated the Status to the stakeholders**. It is also important that you have received the agreement to move on.

✓ **Presented the Findings in some fashion**, even if it is just a summary document or memo. Of course, a formal presentation is great but remember, it is not just you who should fully comprehend, but everyone involved.

Once you have checked these off your list, you can move on, with confidence.

Part 2: Plan

Chapter 6: The *Plan* Phase

Remember, our methodology has four phases:

1. *Understand*
2. *Plan*
3. *Change*
4. *Maintain*

Understand Leads to Plan

Now that we have covered the *Understand* portion, our next concept is the *Plan* phase. This is a topic I used to simply call *Design*, until I realized a "design" is not the actual result.

You are coming up with an actionable plan that includes the design as well as the project planning—identifying *who*, *what*, *when*, and *where*—and recognizing the *why* identified in the *Understand* phase.

My favorite way to think about the *Plan* phase is to remember my time in drafting classes in high school. I would sit at an oddly angled table with graph paper, a special pencil, and other strange tools, and attempt to draw plans for houses (floor plans, elevation views, and even plumbing blueprints, at one point). In a way, what we are doing with any design or *Plan* phase is laying out what success looks like in a way that multiple teams can use.

Often, there is a temptation to jump straight from *Understand* to making a *Change* (or changes), especially when you know what must be done. However, you need to be patient. Assess the weight of the importance and risk associated with it. Jumping straight to Change is a bad practice.

Think of it this way: if I were to have a floor plan drawn up and the people installing the drywall saw that the framing was completed, they may assume that it would be fine to start their work and get their tasks done ahead of schedule, simply to make an already angry boss happy. But what if the electrical work had not yet been considered? The work would have to be re-done as you can't have electrical wiring outside of the walls! A well-formed *Plan* process will prevent those kinds of things from happening.

The *Plan* phase seeks to outline all changes that will be made and to document the intended process. We recognize here that last-minute adjustments will likely happen during the *Change* phase, but our goal is to come to an agreement of all considerations ahead of making those changes, to prevent risk. Preventing risk is what this whole Methodology is about.

The most important output of a *Plan* phase is the written design, which can take several forms. Here are a few examples:

- List of Software updates to a defined group of machines with an implementation plan

- Conceptual Design awaiting Proof of Concept Validation
- Validated Design
- Disaster Recovery Plan
- Service Desk Operational Services Plan/Design (support workflows, etc.)
- Anti-Malware exceptions list
- Issue Resolution Plan

The main thing to remember is that you are, in all cases, communicating what we will commonly refer to as "design decisions." Your design decisions will lend authority to our plans and form reference points for the future.

Design Decisions

We make decisions rapidly. It is how our brain, the ultimate differencing engine, works. It is often easy to take for granted, or just forget altogether, that others don't realize the decisions you have made. For example, this morning, instead of driving my BMW sports coupe, which my wife generally expects me to drive, I decided to drive our Hybrid SUV, because I knew we needed to get the oil changed in it. My wife may not realize why I took that vehicle, unless I tell her about it. This is more or less what we are attempting to do with communicating design decisions in our plan: tell others what is going on in our minds and *why* it makes sense to take the actions we intend to.

Design decisions are questions that are answered in advance and determine what will be changed in the next phase.

Some design decisions are rather large and have several implications. For example, what hypervisor or Cloud provider should we select for our workloads? Until you know more about some other factors, you may not be able to make that determination. For this reason, I recommend starting with the items you know you can "fill in the blanks" about, and work out the others as you go along. This may sometimes seem out of order because it often is. Smaller design decisions may be as simple as what to click or type into boxes when you are configuring software. It may be the number of virtual CPUs a virtual machine requires. No matter how mundane, they need to be recorded in order to reduce the risk during the *Change* phase. This will also reduce the amount of time spent in asking questions "on the fly," because they have already been answered.

Even decisions about updates to systems should be cataloged. What updates will be applied and why? Who will support the changes once they are made?

Thinking Ahead

A key component of a successful conceptual or detailed design is simply thinking ahead.

You should think in detail about what should be and will be put into a documented format. This is especially true in Infrastructure Designs. The modern IT world does not do well with the "box" or "all-in-one" solution because it does not scale well and is often inefficient. Modern IT uses shared components and services for maximum efficiency.

So, what happens when needs change? What happens when more users will be accessing the system?

These are the kinds of questions that are addressed in the Design portion of the *Plan* phase.

I used to have an unfortunate curse that I've turned into a gift (or superpower!). I will often sit and think about what might happen and work through different scenarios in my mind. It almost feels like looking into the future.

However, all I am doing is drawing on past experiences, so I can complete a timeline. Although it has not always been productive in my personal life, I will tell you that it has been productive professionally.

If you can't confidently answer the questions that come up in this process, the best thing to do is to re-engage the teams you had during the Assessment (or whomever is appropriate) and ask some leading and open-ended questions, such as the following:

- In what ways are you anticipating growth over the next year? How about the next five years?
- What would the impact be of growth you could not accommodate?
- How often do you require software updates?
- How much has the usage of the software changed over the past year or so?
- What things are users asking about or complaining about?
- What other initiatives are happening in the next few years that could impact our Compute layer capacity?
- How will we be training users or support staff?

Something I want you to notice about the above list: they are all *who*, *what*, *how*, and *why* kinds of questions. The amusing thing is that they just naturally came out that way. This is because I have trained myself to think in terms of open-ended questions and I highly encourage you to do the same.

Thinking of open-ended questions leads to even more forward-looking thoughts—not of doom and gloom of what may happen, but of what could be done proactively to prevent or reduce risk.

You will also need to think in terms of your experiences and borrow from the experiences of others—just as in the *Understand* phase. Make sure you educate yourself on what things may (or should) occur! In addition, keep mindful of market trends. You may find that what used to be a leading practice last month has changed, due to conditions in or around the software, political climates, or the current state of worldly affairs. Keep a network of information and contacts at your disposal so you can see what works for others. From there, you can set up conceptual designs based on what most commonly works.

What if your plan is only for something simple, like installing updates to an already running and stable system? You're not exempt! In fact, I'm willing to bet that over half of the people reading this will have Change Control procedures to which they are required to adhere. Undoubtedly, these procedures require you to have a back-out plan for any changes. Why? Because you need to think ahead, not only to what will be needed in the future, but how you may handle scenarios where things don't go according to your *Plan*!

This means that part of your process needs to be thinking about procedures which you may not be responsible for at all, but which may affect you. Having backups of data, plans for failover, and even for handling support calls should all be thought through. Outline who will be responsible to ensure each part is in place and discuss everything with your team.

My bet is that you will uncover a lot of needs in this process. You will need to write them down. As we did in our *Understand* phase, we will be outlining the next steps and recommendations in our documentation after a review. Remember that this methodology is an iterative process, so we are always looking for what will happen in other phases.

Development Lab

"There have always been ghosts in the machine..." —Dr. Alfred Lanning, the character in the movie *I, Robot*

There will always be ghosts in the machine. Naturally, just like when you are building a home or office building, you will need to make adjustments to your plan. You will also likely find yourself having to make some decisions you did not anticipate. Although you cannot know or anticipate everything, you should shoot for having the vast majority of your needs covered in your design.

A common question when it comes to the documenting of the design, then, is "How will I know what design decisions I will have to make?"

The answer for this varies. Ultimately, if the technology is not something you have worked with enough to have a template built with design decisions to be made, then you will need to create one. This typically means doing development (Dev) work—sometimes referred to as having a "Lab" or laboratory environment that allows you to simulate conditions enough to be able to document the design decisions you believe will be best suited for your project.

In school, a lab was used to learn through experimentation. Your IT lab should have the same mindset, which is why I will typically use "Lab" over "Dev." The main difference, conceptually, is that a Lab is for learning and Dev is for working toward specific goals. Therefore, many companies have both.

A Lab or Dev environment is used to reduce the risks of damaging production while still allowing more flexibility than a Test environment. We'll talk more about this in the *Change* phase, when we discuss Testing Procedures, but here is a quick sample of workflows and risks.

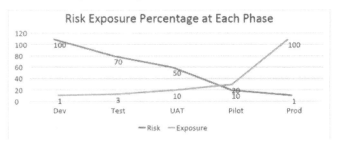

With each level of flexibility, you will note that the risk of affecting live data, production users, or customers increases. Because we do not want anything to be able to hold us back in making changes (no Change Control should ever be required for a lab), it is important that our Lab/Dev environment be free of as many production exposure risks as possible. You are making changes that could impact any number of systems of which you may not be aware. You should take steps to isolate as many components as possible. Database servers, Directory services (like Active Directory), and even virtual and physical networking and storage should be isolated, if possible.

Here are some criteria a Lab or Dev environment should always have:

- No Production Workloads
- No Production Data
- No Production Users (this will be done later, during validation and testing)
- Does not "become" production (this is for learning procedures and repeating them for production)
- Is not concerned with performance, which is a good way to use commodity or older hardware
- Freedom to be changed, at any point
- Share as few components with production as possible
- Fully destructible

There are a few optional, but recommended, aspects of a Lab or Dev environment:

- Does not consume licenses. Most vendors offer either development or trial licenses. I recommend utilizing them whenever possible.
- Completely isolated from production. By "completely," I mean just that—a fully safe Dev environment shares no components with production, except perhaps power. This will give the best autonomy possible.
- Run in a dedicated virtual space. I personally run Lab environments from my laptop and main PCs that have a router completely separating them from the rest of the network.

Briefly, I want to touch upon a very bad habit I have seen based on a misunderstanding of how DevOps works, which I feel presents very real risks, operationally. When I say your development environment should not become production in any way, there are very good reasons for this. These reasons are:

- You will eventually need to upgrade… everything. Locking yourself into a specific configuration of any kind by linking Lab (Dev) components into production can be catastrophic. This absolutely includes considering the Cloud, as upgrades to Cloud infrastructures are not as automatic as some people tend to believe. In a similar way, rebuilding your entire lab every time is fairly impractical.
- You will become more confident in trying new ways of doing things, if you have a dedicated space in which to do them with no time constraints.
- You need to be free to make mistakes—mistakes that could very well carry over into production, if the DevOps model is followed from non-production stages too early.

Perceptions matter. Users exposed to a volatile environment can "poison" the perception of the project early on.

Chapter 7: Dependencies and the Change Plan

Another area which the Architect can reduce risks is in defining dependencies for the *Plan*.

Although the Engineer will typically make suggestions in this area, ultimately someone will need to review the needs and design decisions, then create a list of dependencies. These dependencies will determine the order in which each process can be completed. Because multiple initiatives will often be going on at any given time, communication between departments or focus areas is very important to accomplish this goal. Just as we discussed in the Thinking Ahead topic, leaders in the project will need to think about each dependency and put them into a timeline for completion. This will be called a Deployment Plan or—more appropriate to our methodology—a Change Plan.

For example, if we have a project to introduce Multi-Factor Authentication using a third party provider, the success criteria is that the users will need to use this third party system to log in to their web-based applications as well as to their VDI desktops. Working backwards from the VDI desktop, we know that authentication must contain a token from the domain, so the Architect has included Federation Services from Citrix to generate a virtual Smart Card identity. This Federation is occurring from a Citrix Gateway (SSL Proxy on the ADC), which is utilizing a Security Assertion Markup Language (SAML)-based third party provider called Okta. Okta uses an agent to communicate with the local Active Directory to synchronize identity and tokens.

The dependencies are:

- The VDI must be of a certain version
- Federation Services must be configured, which requires very specific Domain Certificate Services access
- The Citrix ADCs must be of a certain version
- The Active Directory must be at a certain level

Which do you tackle first?

1. We determine that the Active Directory must be the first upgrade to support Certificate Services and ADFS (Active Directory Federation Services) for SAML, which depends on another team. So, we place that dependency first.
2. We then must configure the SAML provider.
3. We configure the Certificate Services (CA) on the domain.
4. We see that the VDI system must be upgraded so that team must complete their actions next.
5. Federation Services and the Citrix StoreFront are upgraded.
6. Citrix ADCs follow, and then testing begins

Congratulations. You now have the start of a *Plan*!

Hopefully, the idea is clear, while only one goal was in mind, multiple components depend on each other, so a proper timeline must be defined. Timelines will affect Project Management in a very big way, so it is crucial this step is not overlooked—even for something as routine as applying system updates. A clear understanding of *who* will be doing the updates and *when* is important.

Different past experiences and stories come to mind, especially ones of conducting updates to a storage system in the middle of a database upgrade, which corrupted the database. One story, in particular, also happened to include a failed backup that a team did not find before starting, so the upgrade could not be backed out, causing a very bad next day (actually, it was a very bad three weeks, if I remember the story correctly). If communications around the timeline of events had occurred, and if the dependencies been properly recognized, the corruption would not have occurred in the first place. Additionally, if the team verified they had a good backup, that dependency would have saved them from making any changes at all and they could have aborted.

The bottom line is, think ahead about what you will need for the changes before they are happening and you will reduce risks.

Once you have determined the *Plan* for changes, the next step is to outline validation. Earlier, in the *Understand* phase, we defined our Success and Testing Criteria and drafted a Testing Plan. That plan should be updated and agreed upon, given the new plan. The Testing Plan should be integrated into the *Change* process for each individual Success Criteria. Define who will be responsible for conducting the tests, who will give approval that the criteria have been met to satisfaction, and what the next steps will be after testing (for example, creating a how-to document for users).

Chapter 8: Plan Documentation Content

I commonly get the question, "Is a spreadsheet good enough for a Design Document?" and the answer is the best of all Consulting answers: "It depends." As in, it depends on your procedures, customers, and teams (AKA your audience). My advice is to consider project stakeholders and verify the style of documentation before you begin. I can tell you that more than once I have encountered a project where the assumption was made that a spreadsheet with design decisions would be enough, only to find that the company required text documentation for compliance purposes.

Some other considerations are to include the overall goal of the *Plan* and how it will be used going forward. For example, if the only goal of the *Plan* is to perform a single specific change, such as a software update, the only thing that is likely needed is an update to the existing document. In that case, you will want to treat the document as a "living document," where changes to the document are expected as changes are made. In cases like that, a spreadsheet tends to work well.

On the other hand, if you are a consultant who is building a system for a client, having a "sealed" point in time for your document may be important, so delivering via an Adobe PDF or locked Word document will be important. Or perhaps the only need for documentation lives in an electronic format such as a Change Control system or SharePoint site.

Regardless of the type of documentation, the most important aspects are that documentation be:

- Standardized
- Accessible
- Consistent
- Appropriately Comprehensive

Document Types

Now that we know the documentation format need, we will need to determine the type of document we will be making. To get us started, we will look at larger projects, design decision tables, diagrams, and the living design document.

Most larger projects should be considered to have an in-process documentation style—where the document morphs, expands, and improves with each pass of the iterative process. At each update, risk is reduced and confidence in the intended configuration increases.

Conceptual Design. This is the most common documentation for design decisions that have not yet been validated or tested in production. At times, this document may be used in conducting a Proof of Concept or Validation test prior to a formal design process. The Conceptual Design should be careful in not using definitive wording—all design decisions are essentially best intentions at this point, until they are discussed or validated. Use words like "intended" or "should consider," at this phase. Conceptual Designs are often composed as part of a pre-sales process if you are a Professional Services organization that also sells products. Either way, I recommend creating a Conceptual Design after an Assessment and prior to conducting *Plan* meetings, even if the Conceptual Design is little more than a diagram of the proposed environment. The goal of a Conceptual Design should always be to be converted into a final or validated form. However, be aware that some companies will ask you to build a Proof of Concept at this phase, which we will cover in the Testing Procedures chapter (Chapter 14). The risk here is that the design is conceptual, so it should never be exposed to production data and/or users. A Conceptual Design is not adequate for production use, nor is a Proof of Concept typically appropriate to place into production.

Proposed Plan. From our conceptual process, you would typically now conduct interactive sessions with technical teams to ensure that the design meets the success criteria. Once agreed upon, the design decisions are written down for the test/validation and production environments. These decisions are often reviewed by an architectural team and/or presented to project leadership prior to a commitment to build. Our language changes to a "will build" style of language. At times, the Proposed Plan will include Build Materials (or Bill of Materials—BOM), listing the resources to be consumed, purchased, or otherwise acquired for the project's success. This important aspect gives management the ability to anticipate costs. The goal of the Proposed Plan should always be to lead into an Alpha/Test build in the *Change* phase.

Final/Validated Design and Plan. Highly recommended in our Professional Services Methodology is a process of composing a conceptual Design and Plan, conducting a validation build (either as a Proof of Concept or Test environment), understanding success, adjusting design decisions, re-validating, and re-assessing. At the end of this process, we are much more confident of what will be built into production for your Beta/User Acceptance Testing and your Pre-Production Pilot. This type of document is very commonly converted into an "as built" document once production has been approved.

Operational Support Plan. Sometimes referred to as an Operational Design document or a "runbook," this document contains how the environment will be supported in the *Maintain* phase. While this information may be repeated in the Standard Operating Procedure (SOP) documentation in your organization, I highly recommend this be documented separately first. This will also be a "living" document—updated and expanded to meet whatever needs occur. Conceptually, this document will answer questions such as:

- Who will maintain the environment?
- Who will administer the environment?
- How will the environment's data be backed up?
- What are the Disaster Recovery and Business Continuity Plans?
- How often are updates applied?
- How often are machines rebooted?

- How will the service desk troubleshoot issues?
- When and how will Security be reviewed?
- What compliance plans are required and how often will they be updated?

(Note that we will cover this document in more detail in the *Maintain* phase, later on.)

Environment Design Summary. Especially in larger organizations, other Architects may want you to maintain what the needs of the environment are. Very often, this will include the number of virtual machines, storage requirements, CPU, Network, and Memory needs. This document does NOT typically contain details about design decisions and should be kept to about one page. Diagrams are encouraged. However, consider that your audience will likely not understand the nuances of what each component does. For the whole of this summary, the consideration should be to keep it to the point, as if you had only five minutes to describe what is required to make your design work.

Design Decision Tables

An effective way to communicate design decisions that is easier to read and can be more quickly referenced is to use tables. I like to use tables largely because I can typically either fill in the "blanks" of a template more easily or utilize a spreadsheet to fill out the information and import that into the document.

Ultimately, tables are something that take practice in making, but the criteria should always be to reduce the amount of reading and especially avoid redundancies wherever possible. Tables should tell the story quickly and in a way that is easily changed. While this is not an absolute rule, I use tables any time I have information that would be used more than once.

For example, tables are useful in describing a Requirements Table. Instead of listing out the individual requirements each time, having a table can rapidly summarize the information.

Each row of a Requirements Table represents example requirements.

Name	IPs	vCPU	vRAM	vHDD
Server 1	10.0.0.10	2	8	80
Server 2	10.0.0.11	2	16	120
Total	2	4	24	200

Or, you may need to have a table for Active Directory policies, where the same policy is linked to multiple OUs (organizational units).

Policy	Infrastructure Servers	Developer VDI	Accounting Desktops	RDSH App Servers
Firewall	no	yes	yes	yes
User Profile Management	no	no	yes	yes
Local Power User	no	yes	no	no
Internet Explorer Rules	yes	yes	yes	yes

Be mindful of fonts, color schemes, and the overall theme. Tables should also be consistent in how they break across pages (allowed or not). If a table does break across pages, be sure to repeat the headers on each new page. I also will typically include a caption for each table, especially if I have allowed the table to break across multiple pages.

The Power of Diagrams

"A picture is worth a thousand words."

—Proverb

Diagrams are one of the most powerful ways to ensure your points are understood. They are valuable in documentation but also in day-to-day working. Gaining skills in diagramming—electronically and on whiteboards—will always pay dividends in your career. Whether you are a Pre-Sales Engineer who is explaining a high-level concept to a potential customer or a seasoned Architect who is trying to bring multiple teams to understanding the implications of the new design, diagrams work when words fail. There are some criteria you should be aware of for your diagrams. As fate has it, they all start with "S." (Okay, fine, I made them all start with "S" on purpose.)

Simplicity. Some diagrams, especially networking detail diagrams, can become so complex that they cannot truly be read or followed.

It is best to keep diagrams as simple as possible, only including relevant information. If your diagram cannot convey the intended results within about 10 seconds of viewing, it may be overly complicated.

However, complicated diagrams are sometimes required. In those cases, I will typically keep those kinds of diagrams for an appendix or deliver them separately with the documentation.

Subject Matter. It is better to split the subject matter of diagrams into multiple diagrams rather than have the "all-in-one" diagram. A key for success here is often to use the same diagram set and modify the interactive lines to describe your subject. For example, if you were diagramming a VDI environment, you may want to have a master outlay for the conceptual environment, then add in networking descriptions, or even use a user workflow diagram.

Style. Visual details matter, and the more consistency you have in your diagrams, the more your subject matter will stand out. Color coding is something I firmly believe should be used, whenever possible. If you are color-blind or color-deficient, don't worry. I'm not recommending at all that you should replace text with color. What I am suggesting is to use color as another way to quickly reference items within your notes, presentations, and documents.

For example, if you are creating a high-level conceptual diagram of the overall design for the beginning of your design document, each conceptual layer can be represented by a color. If you chose red for the User/Subscriber Layer and green for the Access Layer earlier, you can and should use the same colors in any additional diagrams and callouts you make.

Size. This is often missed, but even if you include a well-formatted and beautiful Visio diagram in your document, if the text can't be read from a distance, you have made it worthless to the reader.

If your diagram has too much information, see the aforementioned Simplicity point.

Examples

At times, you may be asked to make compromises on this to explain visually to multiple teams with the same diagram. While not ideal, at those times you must look at your diagram and ask the question: "Could I convey this in a more simple way?" For example, the following diagram was requested to convey the workflow, network traffic, components and infrastructure in one overview diagram. While my team and I did the best we could and delivered the document, we also used several colors, shading and small text that do not print well. While it is effective in conveying the layers and execution in one diagram, if you put this on a screen during a presentation, you would likely get more questions than you were wanting.

[Unfortunately, the nature of the print for this book format does not allow for a great example – however the PDF version is available with your purchase and does contain examples that are useful!]

The Living Design Document

Because we are finding success with the Methodology, we want to encourage that continued success.

One of the best ways to encourage ongoing processes to be followed is to deliver "living" documentation.

This simply means that the document can be updated as time progresses, needs change, and expansion occurs.

There are a few criteria I recommend for this to work well, which are:

Have a Revisions Table. Either at the beginning or the end of a written document there should be a table that lists the details about the overall version of the document (more on that in a moment), what was revised, who revised it, and when. You may want to include additional notes, if required.

Use a Review Process. An important aspect of a document structure that will be updated frequently is to ensure you have a good review process in place. I recommend the following phases:

- Content Development
- Technical Content Review (a review of the technical details and the sign-off from Architect)
- Quality Assurance (grammar, formatting, etc.)

Use Sensible and Consistent Versioning. In software development, you will typically find various sub-version levels depending on the style of development. This is to track both the status and reliability of the document in the same way software would be tracked. For example, you may only change your primary version for major *Plan* or design reviews. For your sub-versions, you should only keep track of a single depth in the document revision table (e.g., 1.2), but in other versioning you may keep multiple other levels, depending on your workflow.

For example, you may have multiple people working on a revision so each person would submit their version of the document for revision, perhaps saved with their initials (e.g., Design-1.2.3-DJE.docx). Ultimately, find what works for you. Just remember that a good practice is to be consistent.

Chapter 9: Documentation

Written documentation is the single best way to prevent risks.

In a fast-moving world, such as Information Technology, it can be very tempting to not slow down enough to write things down for later. We are tempted to deliver things verbally or simply rely on an email. When I have found myself behind in creating several deliverables, documentation suffered. But I will tell you that if there's one thing you want to go back and make sure is done—even after the fact, in some circumstances—it is documentation.

One of the comments or complaints I hear most often from engineers is that they don't feel they are very good at writing. So, they find ways around it, and our fast-moving IT culture allows it to happen. It's a shame because someone will often have a question months later and by then it will have been forgotten.

"How long should my documents be?" I am asked this common question frequently. I have had the unfortunate experience of being part of teams that were told that their documents had to be over 100 pages to be valuable! I beg of you, please do not torture your audience in that way unless they have specifically asked you to provide that much wasted space! More is not always worth more. In fact, it is often worth less because the hours you spent writing extra content could have been spent on another phase of the methodology! So, again, the answer is the most common phrase uttered by consultants, "It depends."

It depends on how much content you need to convey. Ask yourself, *What will be nice to have as a reference in three years? What would I want to see? What wouldn't matter?* That attitude will affect your decision as well as management's expectations. Some documentation is better than none, of course.

"What should I include in my document?" We'll cover those details in this section, but at a philosophical level, to make documentation that is valuable, it will need the following content:

- What
- Why
- How
- Who and When (optional)

You'll note a distinct lack of things like opinions or other means of being overly verbose. (I save that for books and so should you!) Keep to the point by constantly fitting what you are saying against the above simple criteria.

It is best to essentially tell the story of the environment, the changes made, the intended changes, or the *why* of a process. Even if it is a simple Windows Update process, document what needs to happen and why. Believe it or not, it is often a page of text stored in a SharePoint site or something similar that can do wonders in determining what courses of action were taken at a certain time and why. So, if it's a matter of determining what backup to recover, months after the fact you'll be able to do so with confidence.

What if you are not good at writing? There's really only one way to truly get better at writing, which is to write. Find a mentor (or pay one) to review your documentation and give you pointers. That is single-handedly the way I improved, so much so that I still am in the habit of having peers and mentors review my work. Practice is the only way to improve. I'm sure there are courses out there (if not, I'll make a note to start one), but even those will have you practice. So, suck it up and get to writing.

In most cases, the most detailed writing that will be done is actually during the *Understand* phase, but the concepts here will largely continue in each phase.

As you probably guessed, the actual content of the Assessment document will vary greatly based on the project's focus. But in all cases, I recommend an adherence to the Layer approach. This will ensure a level of accountability but also ensure that you are giving the complete picture.

The Importance of the Executive Summary

If someone is paying you to perform a task, you need to show them value. So, for every interaction, no matter how simple or complex, I always recommend an Executive Summary preclude all other document text.

The summary should be just that—a summary. I have heard it said several times—and I agree—that if your Executive Summary is longer than about half a page, expect it to be ignored. Of course, shortening it will take practice. If you are very technical in focus you will likely have to hold back your urge to explain everything. The average executive will ask you when they want more detail. They are busy with many things and typically have people working for them who will be concerned with the details, so they need to know a few key things:

- What risks are there today?
- What resources am I going to need to address these risks?
- Are there things to consider beyond what we are doing today?
- Are the needs of the business being addressed in what we are doing or proposing?
- What timelines and follow-ups are required?

This is to name a few. Getting a feel for what is appropriate here is often more art than science, but those who are successful can quickly get their point across while showing confidence in their work. If you do neither of these things, you may find yourself on the back of the list of promotions and the project on the last of the list of things to be done!

Chapter 10: Risks and Recommendations

In nearly every phase of the methodology, you will find that you need to point out certain risks to the project, company, or others. We talked about this in the *Understand* phase. I also cautioned you to be appropriate of the time in which you point out risks and give recommendations. We will expand this in a moment, but first let us talk more about WHY this is important. In fact, this may be the most crucial aspect of the methodology because every phase will seek to minimize risks. In order to minimize risks, you must first *UNDERSTAND* the risks before you can *PLAN* to address them.

Let me get something out of the way because I see it far too often: "Because I said so" was never good enough for you when you were a child, so why would you ever expect that answer to be adequate for the company you are serving?

Just because you have learned a few things, or are even considered an expert, does not give you a license to give blind advice. Many are tempted to do this because it seems to build one's sense of confidence.

However, that is not a sense of self-confidence; it is one of pride. If a company is only served by having you around all the time to answer questions because you have not enabled them to understand why you have made the directives you have made, then your career has reached as far as it will ultimately go. You are stuck and will be exactly where you are now in five years—burned out and frustrated—because while the company has confidence in you, it has too much confidence and will never let you take a vacation. Sure, the money will probably be good, but you will not progress beyond your current position. And believe me when I tell you that only you can get yourself out… and that there is more opportunity and money you are missing!

I think of an example in the book, *The Phoenix Project,* that points this out well in a true-to-life scenario where the processes of a company's IT department, development, and production were all dependent on a single person. It quickly got to the point where the management recognized that the person was their single point of failure. No one understood the processes well enough to do them on their own, and a very talented person always took it upon themselves to fix things because it would take too long to explain rather than to just fix.

Does that sound familiar?

A link to the Phoenix Project and the audiobook can be found on the Methodology Book Resources page at https://it.justdothis.net/bl#5 and I highly recommend you buy it. Don't borrow it. Buy it and read it more than once. It is an often painful reminder of how so many IT processes remain to this day. If I were there, I would have given some different advice!

First, know the risks and make sure your audience understands them.

Second, make recommendations that enable positive changes *without* depending on you, personally, to complete them.

Making a Recommendation

Be mindful of what you know, and others do not. Be aware that your audience may have a cursory understanding of the technology, but may not fully understand all its implications. When defining risks, you should explore all the aspects of that, then write your understanding of the risk with those aspects—and your audience—in mind. This can sometimes be an exercise for you as well. Very often we do things simply because someone else said so. We don't fully understand it ourselves, but we repeat it because we have it on good authority that it is true. Aside from the obvious risks of that practice (which you may misrepresent or not fully understand), you must be able to walk through in your mind (and sometimes on paper) what will happen if your recommendation is not followed.

For example, at one point I was giving out recommendations to customers that they should follow the non-uniform memory access (NUMA) values of their central processing units (CPUs) to determine the number of virtual CPUs to allocate to a virtual server for their Remote Desktop Session Host (RDSH).

The problem was, I did not fully understand, at the time, the recommendation. So, I would not list that explanation. Instead, I listed in my recommendations that the servers should have three virtual CPUs instead of four. In my mind, this was apparent because the physical CPU was a six-core processor that matched a NUMA value within that processor set. However, this was not apparent in their minds!

When I visited that client for a follow-up a year later, I found that they had been deploying servers with three virtual CPUs, as per my recommendations, but that performance was not as it should have been. By investigating, I found that they had, in fact, upgraded the processors to new CPUs with 14 cores. Since 14 is not directly divisible by three, performance was not as it should have been. But because I had not fully understood to explain the *why* in my recommendation, they were following blindly, not understanding the implications. They just thought that an expert had said to always use three instead of four! I appreciated their faith in me, but was glad to educate them fully, after the fact.

Choose your wording carefully. You are only likely to have your recommendations followed if you effectively communicate their importance. If you are giving a well-formed risk, do not follow it up with a weakly-formatted recommendation.

Be authoritative when you are giving actions to follow. I see two very common mistakes here, overconfidence and timidity. Both can cause your audience to dismiss your recommendation. I often fall into a trap in this area of not being authoritative with my wording. In conversation, I often struggle with the thought that I am the authority. I'll revert to passive language, which causes the recommendation to lose power. Passive language includes "Leading practices are to _____." Or "Microsoft recommends _____." Those seem like good recommendations—and it is easy to think of them that way—but the problem is that they don't include a call to action.

You need to be aware of the following helpful tricks I have learned and put them into practice, which are to:

Use Active Language. You have done the work of determining the actions to be taken. You now need to assign them to the next person(s) of responsibility.

- "Company should configure _____" is active.
- "Company will need to contact Cisco to resolve the issue with _____ before continuing with _____ in September" is active with a timeframe.
- "Company has decided to engage consulting to complete the task of _____" is another example when the action steps have already been decided.

Be Aware of Personal Prejudice. Going into a recommendation, you may know that a company's staff doesn't have the will or perhaps the expertise to complete what you are recommending. This can be especially true if the recommendation involves something that is not in the current budget. However, be careful not to let that color your wording of the recommendation.

Keep in mind that you are writing a document that will be viewed more than just once; more than for just the time being. If you are making a recommendation with that in mind, you have done your job already. It should also go without saying that your text should never contain anything that assumes or pre-judges anyone's motives, religion, or racial factors.

Don't Oversell it. "Strongly recommend" or "strongly consider" are examples of wording that is not always appropriate. You are making the recommendation already. If your description of the risk was not strong enough for them to know the importance, now is not the time.

Give the Information Generously. Earlier in the process you should have done some research. You should absolutely feel free to include links and statements similar to "For additional information, refer to…" in your recommendations. Web links should either be cataloged at the end of a document, in a footer, or even in the text if the URL is short enough. However, I will caution you that if you are giving a web link, you should include a searchable term with the URL printed out along with it. Do not assume that the format the reader has the document in or is viewing the document with will include clickable links.

For example, a link sentence would look like: "For additional information, see the Methodology Book Resources site at https://url.com/resources-are-over-here/link.html." Or "For more information, please see the Resource Links section in Appendix A." For the reader's convenience, go ahead and link the text in the event they are viewing electronically.

Be Aware of Length. As you can see from the previous points, you are already giving a lot of information in a risk and recommendation statement. Keep in mind that these paragraphs tend to be targeted towards executives. Therefore, keep it as short as it needs to be to convey your action items and *why*.

As aforementioned, if additional reading is required, link or direct the reader to the text of the document or the appropriate other place to find the information.

The Risks and Recommendations section should remain as brief as possible and easy to read.

The Magic Paragraph

Earlier, I talked about consistency in your message.

When I had been consulting for almost a year, I found I was struggling with this. Documents I would send for review would regularly come back doused with comments to address. This was especially true in my Assessment documents because I did not yet have a clearly defined message that was consistent, regardless of the situation.

Finally, after some head-butting on the topic, I finally asked my Service Delivery Manager from Citrix Consulting (at the time that person was Scott Campbell—and I attribute the following to him because I don't know any different sources) what he wanted to see from me. What he conveyed to me is what I call the Magic Paragraph.

The Magic Paragraph looks like this:

> **Name of the Problem**. What they are doing now. Why this is wrong. What they should consider instead. Research or links regarding the problem.

In all, this is fairly straightforward. Put in bold type a brief name for the problem. Then, without emotional attachment, state what is being done now. Describe why this is not a correct practice or how it deviates from acceptable leading practices. Next, add value by suggesting a solution or direction of investigation. Finally, include information that the reader can reference as to what can be done, identifying leading practices and other options, for more information. Repeat this process with a new paragraph for each Strategic or Tactical observation.

Let me give you a couple of examples of the Magic Paragraph strategy:

Slow User Logins. Company ABC is experiencing average login times to their CVAD VDI Desktop environment of over two minutes. Industry average logins are approximately 20 to 30 seconds. Company ABC needs to conduct a detailed review of the login process and determine what portions of the process are taking longer than average (for example, the application of personalization or group policy changes). Tools such as the Citrix Director or data gathered in the monitoring tools will typically reveal this information.

Citrix ADC Firmware is Vulnerable. Company ABC is running version 11.0 of the Citrix ADC Firmware. This version is vulnerable to a potential attack that could theoretically compromise the management interface and allow an attacker to gain access to the network. Company ABC needs to update to the firmware versions at or better than those described in the Citrix Support article CTX227928, which describes more about the specific issue. Additional recommendations can be found in the Access layer section.

Notice again, these simply state the facts. Be very careful not to call out teams that have conducted work previously or include philosophical differences of opinion. Simply lay out what needs to change and *why*. Keep it brief.

Speaking of keeping it brief, an important note here is that this is not called the Magic Page or Magic Section. The focus should be on concise information that can be easily consumed by people not involved in the day-to-day operating of the system or who may not know how to describe the problem to others. You will be providing much more detail for most of these observations later in your document or process. Remember, we are trying to make this something that anyone can read with the purpose that they rapidly understand your messages. A professional differentiator is to include a link to the section where you have more information. For example, "A full description of this issue can be found in the User Layer section of this document" and for the words "User Layer," you would create a link to the appropriate section where more information can be found—providing that they are viewing your Magic Paragraph on a device that allows them to click a link.

An important thing to know is *when* to use this format. It depends on your project and audience, but whenever you are writing up any recommendations or pointing out risks, I recommend using this format or a table that is similar.

> ➢ Some IT-only-oriented projects may be much better served with making recommendations in the table format, which we will cover next.
> ➢ Some projects may dictate a combination approach, where the Strategic and Tactical recommendations below the Executive Summary

include the written-out text for the key observations, with the rest made individually, via a table in each Layer section of the document.

➢ Others may be best served by splitting recommendations into Layers only with a Risks Table at the beginning of the document and the Magic Paragraph as the referenced detail later in the document.

➢ It honestly depends on the style you develop and what will allow the best product to go to the audience in the briefest amount of time.

A table that mimics the Magic Paragraph but adds some ability to be sorted or given in a spreadsheet may look something like a spreadsheet template that is included in the resource kit for this book.

Register at https://it.justdothis.net/bl#6.

Chapter 11: Documentation Structure

I have spent time describing the *what,* so now it is time to get into *how* to go about putting this altogether.

An important thing to remember with any documentation is that a good structure allows more flexibility and better collaboration. If you can remain consistent and professional, it will provide a better experience for your readers.

Create a Document Template

I mentioned before that you should not use previous work as a template. I have more than once been part of post-project questions (and at times, during a review session) where the team was asked by the customer "What does this mean? I don't think we even have a (insert component here)."

Save yourself the embarrassment and potential for confusion. Use an empty template without prior or other customer's information. You will get busy and won't always find these mistakes.

Use a Document Style

A document should have a style. This means you should have a specific look, feel, font, and color structure that is specific to your practice and is consistent across all documentation. Microsoft Word has a lot of features you can use to make this process more streamlined. It is worth the time investing in the knowledge to use this properly. This becomes especially true when you start pasting content into your deliverables. In the long run, the ability to merge something into your theme makes things a lot easier.

Sections

In some cases, you may split the document into three sections, depending on the length. For detailed Infrastructure Assessments, for example, I have often put the Executive Summary, Risks and Recommendations, and Environment Overview into the first section. The next would be the Layers, and a final section for any appendices, links, or additional information.

Headings

Another key here is to properly utilize document headings. I typically recommend trying not to go more than 4 headings deep.

However, the ability to have a good Table of Contents and active links between sections makes this feasible and will make your document stand out. More importantly, when your document is built to be more easily navigated, it will provide a better overall value. Headings are supported in just about every major word processor (Even LibreOffice and Google Docs). In most cases, you will want to use Heading 1 (or Level 1) for each Layer or topic. For example, your Level 1 headings in a typical document would be:

- Executive Summary
- Environment Overview
- Business Layer
- User Layer
- Access Layer
- Resource Layer
- Security Layer
- Control Layer
- Platform/Hardware (or Cloud) Layer
- Operations Layer

You also need additional headings under each layer, depending on your content. This makes it easy to keep things organized and makes it easy for teams to divide up work, when it is relevant. For example, you may have seen a structure like this in a Citrix VDI Design in 2017:

- Executive Summary
- Environment Overview
  - Summary
  - Diagram
- Business Layer
- User Layer
  - Use Case 1
    - User Hardware
    - User Access
    - Application Requirements
    - Persona Requirements
  - Use Case 2

- User Hardware
- User Access
- Application Requirements
- Persona Requirements
- Access Layer
  - WAN Networking
  - Access Controllers
    - DMZ Citrix ADC
    - Internal Citrix ADC
  - Citrix StoreFront
- Resource Layer
  - Desktop Composition
    - Use Case 1
    - Use Case 2
  - Applications
  - Optimizations
- Security Layer
  - Antivirus Requirements
  - Compliance
    - PCI
    - HIPAA
  - Desktop Whitelisting
- Control Layer
  - Active Directory
    - Structure
    - OUs
    - Policies
  - Delivery Controllers
  - Cloud Connectors
  - Image Management
- Hardware (or Cloud/Compute) Layer
  - Hypervisor Structure
    - Resource Requirements Summary

- Control Requirements Summary
  - Storage Controllers
  - Networking
    - DNS
    - DHCP
    - PXE
    - Port Listing and Firewall Requirements
- Operations Layer
  - Delegated Administration
  - Monitoring
    - Citrix Director
    - Citrix ADC/MAS
    - 3rd Party Proactive Monitoring
  - Service Desk
  - Image Management
  - Backups
  - Business Continuity
  - Disaster Recovery

You may, in some cases, go deeper in levels, but I do not recommend including levels deeper than three in the Table of Contents. The more organized you are, the easier it will be for your readers to understand—and the simpler, the better.

Tables

Another key part of many modern documents is the use of tables.

Tables are a great way to visually save space. I recommend using tables to outline requirements, design decisions, and configuration requirements (e.g., CPU,

Memory, Disk, etc.). If you have policies or other repetitive information, tables tend to be the best choice.

Additional Document Tips

- Headings can be collapsed in Microsoft Word, which allows easy moving, deleting, and copying from within the document and pasting to other documents.
- Headers should be listed in the Table of Contents. Use an automatic table of contents rather than trying to create one manually. Just don't forget to update the table of contents right before publishing.
- Headers are navigable in formats like Adobe PDF.
- Microsoft Word has a "Navigation" view that allows you to quickly move from one heading to the next.
- Be consistent in spacing, the use of periods, table styles, and fonts.

Chapter 12: Finishing the Plan Documents

Collaboration

Regardless of your document's contents, you should always strive to not be the only person who reviews the document prior to sending it to stakeholders. Whenever possible, a project should have a Mentor—an Architect, Consultant, Principal, or Senior-level person—available to give you feedback. In some cases, it makes sense for a non-technical person to review your work for readability. Ultimately, you will need someone directly experienced to review it with you. Even if they are similar in your level of knowledge, have someone help you.

Because I am recommending Microsoft Word, I will also advise you to use the functionality for collaboration the program has. Use comments over text highlights because they are a powerful way to ask questions and get feedback from others without the risk of leftover text in the document itself. Comments are typically and automatically tagged with the name of the person commenting, allowing you to reply to others' comments individually without having to create a new comment.

Tracking Changes

Another key feature to look for in a professional word processor is the ability to track changes. This can be a fantastic learning tool in addition to keeping changes organized. The Track Changes feature must be turned on. Once it is enabled, any Track Changes feature allows either a "markup" view or an in-line changes view that describes the change in the margin. Each change can be approved individually, so you can start at the beginning of the document and see each change your mentor has made. Of course, if a change has been made that you do not agree with, simply reject the change.

Collaborative Mindset

A key element of all of this is your mindset. If you feel your work should reflect on you well, it is in your best interest to have someone review it to make sure it looks correct before sending it onto management. Trust me, I feel the pain of having to ask people for a review. You feel vulnerable. You worry that people you look up to may look at differently, with a negative connotation, thinking less of you or as not knowing your stuff.

Let me be blunt and personal here: you are not alone. Those thoughts kept this book from being written for more than five years. Seriously! In fact, the rough draft was finished in 2017. It sat, shelved, for two years, because I felt overly vulnerable writing it.

I feel that way typing this sentence.

I resisted the notion of having an editor go over it, pointing out all the flaws. I resisted giving it to beta readers. But I

KNOW people will benefit from this book. Connecting with the thought of hundreds of people advancing their IT careers with intention, making four or five times of their current incomes (yes, it is possible!), and most importantly, having more time at home is a very strong WHY I have finally decided to release it to the world.

Your mindset should be connected with your WHY as well. Why is this document important? What good will come from people reading and understanding it? If it is that important, you must be willing to do what it takes to deliver a quality product. Involve others.

Editing

I mentioned at the beginning of this chapter that your document target should be brief and only become as long as it absolutely must be. Remembering the phrase, "keep it as short as a memo," might help. To give you an example, my process for writing books differs in my process for writing deliverables. For a deliverable document, I will dump as much information down as I can think of. Then I will strive to make it at least 20 to 30 percent shorter. First, I will remove redundancies. This happens frequently when collaborating with other team members. Other times, you may have the same finding in multiple sections. Pick the most important section in which to include it and remove the redundancies, linking to the primary recommendation. I will then see if I can say the same thing more powerfully with a simple statement—or ask an editor to do the same.

Make a table of recommendations or use the spreadsheet resource at https://it.justdothis.net/bl#6 to insert into your document. The process of doing so will help you identify redundancies as well as guide your documentation on what should be given more attention… or less! Be aware of reader fatigue. That is also why I rely so heavily on headers. It makes it easy to reference but also reminds the reader that they don't need to read everything to get back to the guidance they need.

Grammar and spellcheck are a challenge in IT Systems' documentation. At the same time, you're being compensated quite well for what you are writing. Take that seriously and invest in tools to help your communication. In 2019, I began using a software plugin called Grammarly, which has been helpful for me. Sometimes, I will also read the document aloud to myself or record it. If it sounds right out loud, there is a higher chance it is written correctly. Even if you do all these things, it is still best to find someone to review your document after you have done a first pass yourself, checking spelling and doing your best with the grammar.

The flow of a document is another important consideration. In the case of Assessment documents, you will want your flow to address the most important tactical recommendations first. There is a power in convincing your reader of what is important and then guiding them to which actions should be taken first. In many cases, this will also have a psychological effect on the team as they "check off" wins in the document. Wins build on each other. Guide people towards wins and your name will be right there at the top just from being intentional with a document's flow.

In a Design document, you may want to consider a similar flow within each section and start with the early wins, if you can. For example, if you have three primary desktop groups in your VDI rollout, starting with the easiest to deploy makes the process easier to see as possible.

Finally, remember to start with *why*. Your documents have power when you inform first and then direct actions. When you are in the editing process, keep this in mind, and move things around as you need to.

Now that you have completed the *Understand* and *Plan* phases, it is time to move on to the next phase: *Change*.

Part 3: Change

Chapter 13: The *Change* Phase

"Leave it better than you found it."

—Robert Baden-Powell

Remember, our methodology has four phases:

1. *Understand*
2. *Plan*
3. *Change*
4. *Maintain*

Now that we have covered the *Understand* phase and the *Plan* phase, it's time to dive into the *Change* phase.

The goal of *Change* should always be to do exactly what the title suggests: change something for the better. Whether you are doing a small update, conducting a Validation build, or making changes in production, *Change* is the part of the methodology most people know the best. Our goal with the methodology is to reduce our risks during the *Change* phase as much as possible, to give the company and users the best experience we can.

In the past, for consulting projects, I have seen this limited to "Deploy" or "Build" as a title for this phase. I have chosen "*Change*" because I wanted to widen the coverage of the concept as well as simplify the communication of what is being done.

After all, change occurs whenever a target is acted upon to produce a different configuration than it had before the process began.

Regardless of whether you are changing a document or a datacenter configuration, the same concepts apply:

- *Change* carefully follows the plan laid out in the *Plan* phase.

- A *Change* process will typically need to start with a meeting to agree on WHO will be responsible for each *Plan* component or design decision outlined. Another important aspect is to recognize dependencies and the deployment plan. Review and agree to this plan before starting the change; make sure that dependencies are checked and reviewed against the timing of the project.

In order to affect positive change without users experiencing negative effects, risk management is the key of which everyone should be aware. Management needs to be aware that rushing into a *Change* process ahead of the agreed *Plan* is not appropriate, regardless of any desires to have a project completed sooner. Likewise, Engineers should not "cut corners" or "take shortcuts" to get results so they can lessen their workload. It is never a good idea to cut corners or take shortcuts! Therefore, if we are building an environment to conduct validation testing with IT only, it is not appropriate to assume at the conclusion of the build that we can simply move the same system into production. The Architect would not have made design decisions based on production workloads for a validation environment, and taking the "shortcut" could put the project at risk. So, before we get into the typical *who* and *when*, the Testing Procedures of our Professional Services Methodology should be considered.

Much like doctors take the Hippocratic Oath to "First, do no harm," I suggest that you embark on a similar mission. I do this, in all that I do. In fact, my reason for writing this book is to change the way people feel about doing this sort of work. Simply put, I want to leave IT better than I found it. I believe you should have a similar way of thinking when it comes to the *Change* phase. It is important to understand that this phase is one of "make or break." When things go badly during this phase, more people know about it and more problems occur. Having the correct attitude going into it is crucial. If your goal isn't to leave what you are doing better than you found it, you may want to pause and consider your true motivations. If it is just about a paycheck, for example, perhaps day-to-day administration would be a better fit for you. If your mindset does not match the mission, you may find yourself looking for a job sooner than you want to. Mistakes are harder to catch when you don't care!

Chapter 14: Testing Procedures

Testing is the most important and most overlooked differentiator between companies that desire risk management over "just getting it done." In my experience, companies that implement multiple stages of testing have the least amount of negative incidents.

As my friend Jake Hughes, who was the Chief Technical Architect at the Seattle Children's Hospital said, "In six months, no one will remember *when* we went live. They will remember *how* we went live."

Testing before a change goes into production is how you prevent a bad perception, keep management happy (well, happier than they would be, otherwise), and generally avoid those late night phone calls and missed vacations.

Equally important to this is strategically ensuring that users do not have a negative perception of projects, and that management does not see the testing results of a process too early (which could kill progress prematurely).

The testing methodology should include a gradually increasing population exposed to the proposed change, typically in four primary phases plus a production pilot.

When changes are required, you may need to cancel a phase and return to the previous phase. You also may need to do this more than once. This iterative process, much like the methodology itself, should be visualized as a circular flow where changes are possible, right up to the point of production.

Here are the parts of this cycle:

1. Development Testing. The Information Technology professionals (in this case, the Engineers) will take the design/Plan from the Architects and prove functionality in a dedicated, non-production environment. By dedicated, I mean just that—dedicated to that single function of Developmenr. The most successful isolations I see are typically dedicated from the Hardware Layer to the User and Access Layers. This means that compute and storage is dedicated, they have a dedicated Active Directory, their databases are dedicated—everything that is feasible is dedicated. While this is not always possible and you should always check with vendors, most will happily accommodate licensing for non-production use, in this regard. But remember, your audience here is 100% Information Technology teams. The odds of failure are much higher, so controlling perceptions about the project is crucial. The other perspective to have here is that this must be the most agile environment. You must be able to make small adjustments immediately, with only having to inform a small number of people.

2. Validation Testing. Many changes can impact systems in ways that the IT team cannot always anticipate or know. So, the next round of testing will include application owners or perhaps a small group of trained individuals that recognize that they are not to discuss results with other people in the organization but are to give feedback directly to the Engineers or Product/Project Managers, to determine if iterative changes are required. Validation is sometimes required in near-production test environments, but if possible, the IT team should try to avoid exposing production data to this testing phase. The risk has increased because you have typically increased the population of testers to include non-IT individuals. Therefore, fewer changes can take place during this process, simply because having more users makes communication more difficult. Some environments may refer to this style of testing as "Alpha" testing.

3. Proof of Concept. The goal of the Proof of Concept is to seek management and stakeholder approval, based on previously defined Success Criteria. The time frame or how a Proof of Concept (POC) is conducted may vary, but it is best to perform it before User Acceptance Testing (UAT), whenever possible. I like to continue to avoid exposing the environment to production users, if possible, but our goal will be to prove to the stakeholders that the changes have a high likelihood of success in production. If they have not been involved in Validation Testing, I highly encourage you to make sure Service Desk members are involved, at this point. The reason for this is that our next steps will involve users, who will typically engage the Service Desk, if they run into issues. You want to avoid a service incident to be generated because you didn't prep the right teams. Remember, negative perceptions can carry further than you think they will.

4. User Acceptance Testing (UAT). Here, it is time to put the change into a near-production or pre-production environment and ask a pre-defined group of individuals (again, trained to give feedback) to work in the system, typically for a day or so. These people will perform their daily tasks or run a battery of tests. Again, the risk has increased because this group will be larger (about one to two percent of the user population). Remember, even less rapid changes can be done in this phase, because you are asking a larger population to test with their day-to-day workflows. Another goal of the UAT process is to gather feedback on documentation, if needed, to communicate to users about what to expect.

5. Pre-Production Pilot. Now that the changes have been approved by stakeholders, it is time to schedule a rollout to a limited population of users—about 10 to 20 percent of the overall population. In most cases, this lowers the risk of an absolute outage and allows for work to continue, in the event the changes need to be rolled back. I typically recommend between two weeks to two months of Pre-Production Pilot for major changes, but in some cases, this may be as little as a day. It should also be noted that this process is not always possible, especially in certain scenarios involving Healthcare software, which often requires an "all or nothing" change to software all at the same time. Nonetheless, you should always strive to target a smaller population early on, to lower the risk to the business.

6. Production Rollout. Once the pilot is successful and no further changes are required, it is time to make the change for everyone. Make sure your documentation is up to date and that support personnel are prepped. At this step, communication across all teams—not just yours—is very important.

You will likely experience the most stress during the first day of rollout, especially if it is done at night. The more you are ready for it by being properly prepared, the easier it will be.

We will be talking much more about ensuring production stability in the *Maintain* phase.

7. Quality Assurance (QA). Larger or more critical operations may have a dedicated environment that mirrors production with the explicit purpose of crisis mitigation, stress testing, sizing simulations, and load testing. This environment should be updated once the Production Rollout is fully successful. Production users should not be part of the QA environment, as it will typically have a dedicated Active Directory or user database structure, to provide proper isolation for load testing. Very often, a QA environment will be used to validate performance changes prior to rollout into production, or to rapidly test break-fix patches by Administrators charged with the task.

Chapter 15: Change and Project Management Process

The *Change* time frame is the most sensitive time in the project's lifecycle because you are, in effect, disrupting previous methods of operating, which will have attached "drama."

Emotions, fatigue, unmet expectations, and a host of other factors will make this phase the one people remember most.

Given that, just as we have an overall methodology for the entire lifecycle, we need a methodology for how we will manage the *Change* process. In most cases, as of this writing, companies are still using a tried-and-true method called Project Management. In fact, many IT organizations are considered Project Management Organizations (PMOs). This is because they need to be good at adapting to new circumstances and must be organized in such a way to accommodate that.

A newer, emerging trend that is becoming more and more common is the DevOps method. Although it is used more for software development, it is becoming more usable in the IT Services space. Therefore, I have chosen to structure the IT Professional Services Methodology (*Understand, Plan, Change, Maintain*) primarily around a Hybrid Project Management method. In the last few years, I have seen patterns that are increasing my concerns about embracing DevOps—both with untrained individuals as well as with gaps of understanding in teams—and that is why I keep this methodology simple.

The method for the *Change* phase includes primarily identifying plans, assigning team members specific tasks that are based on those plans, and maintaining appropriate communication throughout the process. This method reduces the risks of missing something, duplicating efforts, and making mistakes.

It is important to keep in mind that the vast majority of organizations already have some sort of Project Management system and may have Project Managers assigned to your project, so you will want to work with them and not against them. I'm going to show you how you can do this.

Charter and Testing Plan

At times, a larger project will either deviate from the original scope or have complexity in the *Change* phase emerging from other projects or circumstances, which require all people and teams involved to agree on a course of action, either before or during work being done. When the teams involved are not integrated or are from different organizations, this can cause miscommunication and risk. In such cases, what is needed is a document that outlines mutual goals, time frames, and responsibilities for each step.

The Charter document is a good way to accomplish this without needing to stop work or, in many cases, have a scope adjustment. The Charter typically states who will be doing what work and when. It is very similar to a project plan, but at a more granular level, specifically stating when each team will perform the work. Typically, the document is electronically agreed to by leadership from both organizations/teams. In the case where it modifies the scope, it may need to have signatures and be called a "Scope Addendum."

A common exception would be if the new plan requires additional hours, resources, or licensing; in those cases, a stop may be required as the project goes back into the *Understand* phase to validate why the design needs were not adequate or if the overall plan requires adjustment, given the new circumstances.

Similar to a Charter is the Testing Plan, which should have been laid out in the *Understand* phase—at least, as a draft.

What I typically like to do here is follow one of the habits bestselling author Stephen Covey mentions in his book, *The 7 Habits of Highly Effective People*, which is to "Begin with the end in mind," or look again at my Testing Plan and determine if it will still meet our goals, given all the new information and new plans we have made since we started the process.

As we move along in the building process or what we can call the "micro-phases of *Change*," we can see that the process goes like this:

Change (Build Dev) ---> *Understand* (Does our build meet the Success Criteria?) ---> *Plan* (Identify adjustments required) ---> *Change* (Validation) ---> *Understand* (Test results) ---> *Plan* (if further adjustments are required) ---> *Change* (Pilot)

We need to examine if our testing is properly meeting the criteria and we should also be aware that *who* is testing will be changing throughout each iteration.

Each Testing Plan process will need to define:

- Who is responsible for testing?
- What process is to be followed?
- How will the Testing be performed?
- What results are expected?
- How will results be documented?
- When will Testing occur and for what duration?
- Does the Testing indicate the proper results for the Success Criteria?

Because this will change each time (especially the *who* and *how*), you will need to document the instructions, catalog the results for each testing iteration, and have a Testing Plan for each.

Estimating Time

Lt. Commander Geordi La Forge: *Yeah, well, I told the Captain I'd have this analysis done in an hour.*

Montgomery "Scotty" Scott: *How long will it really take?*

Geordi: *An hour!*

Scotty: *Oh, you didn't tell him how long it would really take, did ya?*

Geord: *Well, of course, I did.*

Scotty: *Oh, laddie. You've got a lot to learn, if you want people to think of you as a miracle worker!*

—*Star Trek: The Next Generation, Season 6, Episode 4: "Relics"*

Easily one of the things that can frustrate me is trying to estimate how long a task or group of tasks will take. Experience is the best teacher, of course. However, if your estimates need to include others, or if these are tasks you have not done before, it is important to be mindful of "padding" your time frames accordingly.

The first thing to realize is the importance of these estimates. Remember, "If you aim at nothing you will hit it every time" (as Zig Ziglar says).

Rarely is "when it's finished" an acceptable (or actionable) time frame. You must recognize that unless you are an IT Department of one, likely someone else will need to know how long to expect to be either without your services, how long a system will be down for a change, or how long it will take because of other dependencies.

Many people struggle with this because of either a lack of experience or other psychological reasons. Quite often, perfectionism will come from the engineering side as well as from the Project Management side—and both are a shame. Often, you may find yourself either thinking like Lt. Commander Geordi La Forge, when you should be thinking like Scotty. Sometimes, you should be acting more like Scotty and less like Geordi! What I mean by that is that you may sometimes be tempted to estimate the time too short, to sound impressive or put people at ease.

But if you don't meet that time frame, it can cause a lot of stress and bigger problems. Other people often depend on knowing when the tasks will be completed. However, if you "under-promise and over-deliver" to be seen as a "miracle worker," you may find yourself seen not only as a bit of a pariah, but you may find that management will dislike you for throwing off their budgetary calculations.

Now that I've convinced you to take it seriously, here are a few pointers to get you started on estimating time:

Keep track of time during each Testing phase. From Proof of Concept, Development, Validation, and User Acceptance Testing processes, you should get a good idea of how much time each task will take and be able to adjust your estimates accordingly.

Understand the risk of variation. Scotty said, "A good engineer is always a wee bit conservative, at least on paper." While you should not overdo it here; you should be mindful of how many outside factors can effect what you are doing. Depending on the specific task, I would advise writing out these factors and assigning a percentage of time to pad your estimate per risk. For example, if I am doing a Microsoft Exchange migration but I know another team is also in the process of an Active Directory modification, I may want to estimate 10% more hours' variance, just in case things do not go well. If you are on a large project with staff dependencies, you should be aware of the odds of staff leaving or being reassigned, which will cause replacements to take "ramp up" time to get acclimated to the project.

Search for advice. I often have to remind myself that rarely am I the only one faced with a challenge. It is why

I have been successful, actually. I received a gem of advice from a college professor that I still keep at the forefront of mind, over 20 years later. It is more important to know how to learn than to learn only to retain information. In our context, this means that it is far more important for you to gain skills in searching Google or some other search engine than it is to try and learn and retain information like this. Likewise, it is far more important to have a network of peers that can help you estimate time and list out tasks, or review what you have and give feedback.

Don't always be Scotty. As I mentioned above, the Montgomery Scott Methodology is not one you should adopt, because it makes for a lot of risk. If you think a project will take 40 hours but you estimate 80 just to make yourself look good, that can really only work a few times. When your time frames are not seen as reliable, time frames will be set for you. Why? It is not just about you; you aren't a miracle worker. You are a professional with a reliable and repeatable methodology. Remember that being reliable will help you get ahead, whereas hoping for miracles will keep you where you are.

Face your fears. This process can be intimidating, especially for independent consultants and engineers without a lot of experience. My advice is simply to examine exactly what you are afraid of and confront it. If you are concerned about missing a time frame and letting others down, you may want to examine if that fear is coming from this experience or other areas of your life. If you fear that you won't be taken seriously unless you can wow people with getting done early, I will encourage you to remember that in the grand scheme of things, there are only two possible outcomes: they will forget in a week, after the accolades have ceased, or they will remember and expect that level of performance, from that point on. If you went through a large amount of stress to meet your timeline, can you consistently keep meet such shortened deadlines? Is it worth it? I bet not! So, you should dismiss that feeling, face your fear, and focus on what you know.

Be mindful of surrounding events. Although most projects are based on calendar time, some are strictly based on time allocation. But even those based on time allocation can still affect the calendar completion date. Because of this, you may be asked to submit how many effort hours a task will require as well as the date when the task will be completed.

Keeping an up-to-date calendar is very important. If you have other commitments, you need to be able to mentally place them along with the tasks you are estimating. In addition, you will need to look ahead for other dependencies and work with project leaders to determine any collisions. For example, be watchful for the following on your calendar and your peers' calendars:

- Holidays
- Meetings
- Vacations
- Company Events
- Other Projects

You may be asked to take your time estimates—along with list of dependencies—and create an Estimated Project Plan. This can be as simple as using a document table, a spreadsheet, Project software, or even online task management software. Regardless, I would recommend having the following information listed for each task:

- Task Name
- Task Number/Code
- Dependencies (by task name or task number)
- Persons Involved
- Hours Estimated
- Estimated Start (calendar day)
- Estimated Completion
- Risk Factors
- Other Notes

Chapter 16: Task Management

Regardless of the project or its size, in order to be effective in the configuration process, the ability to split work into individual tasks is important.

When it comes to task management, it is best to keep things simple, as much as possible. Overly complex systems or getting overly granular with tasks will typically lead to missed updates or even worse, team members who simply do not understand what to do next. As we begin diving into this process, bear in mind the concept of minimally viable efforts. Only put effort into what is valuable! Keeping a process because it "has always been done that way" or because you read it in some fancy book is not a good reason to keep a process that does not serve your current needs.

Project Planning

Typically, this is a collaborative effort between leadership, those on a Project Management or the DevOps leadership panel, and you. Each aspect of the testing and rollout will be given time frame estimates. Project planning should involve the following aspects:

- **Task Assignments.** If the *Change* process involves multiple moving parts or tasks, you may be asked to work with multiple team members. Earlier, in the Estimating Time topic, we discussed outlining tasks and the time required. You may want to assign sub-tasks for each configuration area, based on the skillsets involved. Keep in mind, however, it is often good to challenge staff for growth during these times.

- **Define a Team Lead.** In the event multiple staff members are involved, the team will need a central point of both technical and task authority. Whenever possible, I suggest the Team Lead rotate between peers to train those willing to learn, but the team should always have someone experienced to either take on the role or mentor another in the role. Remember that during the Charter, we should have defined who needs the appropriate communications. The Team Lead will be the person attending meetings and giving updates to leadership.

- **Set Goals and Micro-Goals.** Your overall plan should include larger goals of completion as well as smaller goals—to complete tasks within certain smaller time frames. For example, in an Application Virtualization project, there will be several overall configurations to accomplish, each with several sub-tasks, and each with a person responsible, when possible. The goals in this example might look something like:

Week 1: Prepare for Validation Environment	Person
Allocate Virtual Machines	Tim
Install/Update Operating Systems	Tim
Install Pre-Requisite Software	Sally
Allocate Storage	Tim
Allocate Database	Joe
Configure Active Directory OUs	Sally

Week 2: Successfully Launch WordPad from Validation App Store Internally	Person
Configure Control Components	Sally
Configure Access Components	Larry
Configure Resource Components	Tim
Test Launch of WordPad	Sally

Week 3: Successfully Launch Validation Apps Internally	Person
Install Applications to master image	Tim
Publish Applications	Sally
Configure Persona Management	Sally
Configure Policies	Sally
Test all apps with IT Team	Tim

Week 4: Validate Applications	Person
Coordinate Validation Testing with UAT Group	Judy

Week 4: Validate Applications	Person
Gather Feedback from UAT Testing	Judy
Make Adjustments per feedback	Tim

Hopefully, it is clear that your goals should focus on demonstrable results and not always the tasks themselves. This allows freedom to get ahead, if possible, and also adapt, if the tasks are behind schedule.

Plan to Maintain Visibility. There are a host of tools available that can allow teams to track their projects, but it is important that information never resides with only one team member. "Life happens," as the saying goes. Whether it is an illness or a P1 incident that occurs, your team needs to be able to adapt and take on other people's tasks, if needed. Here are a couple of ideas to get you started:

"Scrum" Boards. This idea from the DevOps way of doing things makes sense in this phase—a brief meeting is held with the team leaders (or at times, the entire team) and a "Who is working on what today?" style of discussion is had. Converse in fashion so you know what people are doing and post the task assignments to a board that everyone can see. I love this for a few key reasons:

- It is adaptable (agile) because tasks can easily move from one person to the next.
- It responds well to changing priorities. For example, a person doing development work can be interrupted to help with a production rollout, to fill in for another team member.

- Everyone can see what is happening and where they can help, even if it is only to encourage their team members.

Project Management Software. Although I don't always recommend it to those just starting out, I feel the industry standard is still Microsoft Project. My main issue with it, in this instance, however, is that the software often lacks an all-team visibility that you ultimately want. I would start with online resources and move onto others. The main concept is that whatever is chosen should ultimately be able to track goals and tasks in a way that is easily understood and be only as complex as it needs to be to accomplish the goal of moving the *Change* phase forward.

Reporting Status

Generally speaking, the *Change* process is what will involve the most coordination of efforts and typically the most time, overall. Because of this and the sensitivity to making configuration differences in multiple systems, the need to communicate with other teams and leadership becomes extremely important. Therefore, we need to have an agreed-upon method for reporting Status.

Most Project Managers will help guide this process by asking you key questions about how things are going, what "roadblocks" or challenges you have encountered, and what the anticipated time to completion will be. These kinds of questions can be frustrating to the person with an engineering mindset; believe me, I have been there. Frustrating or not, however, there is a very good reason for this communication. Project Managers and Coordinators often have central teams reporting on multiple moving components. Knowing when something is ahead or behind schedule affects a great number of other projects, in many cases. Often, there is a level of tolerance for active changes going on at any point (and if there is not already, there should be!) and you should generally be aware of this.

Beyond Project Management, and especially in the absence of it, the need to report Status to leadership is also very important. Leadership may not need to know the technical details about what is going on, but they generally would like to know about what items you have been working on in a specific time period. More importantly, they need to know if there are any problems you are facing that you cannot solve yourself. Typically, these are referred to as "Issues for Management's Attention" because management will need to know about the challenge so they can help you deal with it.

So, during the *Change* phase, regardless of the length of time, you will need to set an agreed-upon practice for reporting Status. I would start with the following criteria and layout for your report and generate it weekly (or at the completion of your tasks, as appropriate):

1. **Date and Time Frame of Report**. (For example: December 9th, Week 4 Status Report for Project X.)

2. **Overview**. This is a one-sentence or two-sentence summary of the project and the people involved in this status report. This should be in "stakeholder" or "manager" language. Keep it simple and to the point. This is NOT the place for technical detail.

3. **Tasks Completed in this Time Frame**. This will be a very high-level (low detail) summary of the tasks completed.

4. **Issues for Management's Attention**. List any issues that leadership will need to help you address. I typically like to use a table for this, to track the Status of each issue.

5. **Task Plan**. This is a table of tasks and each task's status. Here, you will have dependencies called out and responsible parties for each task item. You should also include an ETC (Estimated Time to Completion) or EDC (Estimated Date of Completion).

6. **Hours Consumed Summary**. In many cases, a project will have an overall time frame estimated with effort hours. State how many hours were consumed this week by your team. Include other teams, if asked to do so. Be sure to outline if a project is using too many hours over the estimated time or too many below the estimated time.

Testing, User, and Support HOW-TOs

One of the most common things I see missed in the process of bringing changes live to a group of users is a lack of prior information for said users to understand how to actually use it. What may seem obvious to us is not always obvious to the average user. So, if you are changing the way they are working, in any way (even if it is just cosmetic changes, like the color of a background or icon they are used to seeing), then creating documentation for these users is very important.

This is also a reason I suggest having a "Day in the Life Of" study during the *Understand* phase. Most of the time, in doing such studies, engineers will note that users could be better empowered by simply knowing systems better.

Your task in creating documentation will be to focus on the "what" (what users should expect) and if practical, the "why" (a very high-level explanation, to further empower users).

Support Documents for Initial Testing

Your How-To documentation should walk target users through the process of using the new system in their language and way of working, as much as possible. The document should strive to be as brief as possible, as overly complex instructions can often be ignored or refused. Screenshots and visual representations are very useful—even crucial—for user documentation. Some things to be aware of, however, are:

- Be aware of and avoid colloquialisms or slang.
- Avoid the use of references to color, if at all possible. Instead, refer to a diagram or screenshot with an arrow, circle, or other method, pointing out the step visually in a way that can be identified without a reference to a color. For example, do not have three arrows and ask a user to follow the "green, blue, and then red" arrows in sequence. Use numbers or names to mark your arrows, in those circumstances. This is not just because you may serve those that are color blind; we all learn better this way.
- Always include a way for the users to reach out for help, be it a support number or specific team member to call on. It is never safe to assume a user will simply read your instructions (or anyone's, for that matter) and follow them. Remember, "LIFE happens" to users, too, and changes can be an added stressor.

I typically recommend making documentation in at least some format as early as the Validation step. Creating a basic how-to during that step and asking others to conduct your steps as a point of validation can be very useful to this process.

Recall that we should have Testing Plans outlined. But how will you communicate with your users for User Acceptance Testing, knowing that some of them will not be as technical? The answer is that you will create a User Acceptance Testing How-To document. Encourage User Acceptance Testers to follow your process, even if they know how to do it already. This serves two purposes. One, it allows you to educate the User Acceptance Testers. Two, it allows you to validate how your how-to documentation will be received by users at large. Your Testing How-To should walk your testers through each step of the process they are testing. As previously mentioned, it should also include instructions on how they will give you feedback.

User Documentation and Instruction

Once you have validated both your environment build and the documentation draft, it is time to format for the larger population. Most larger companies will have a communications department you will be working with. In those cases, you may also be required to work with an agent and walk through the process with them. This is why starting with drafts and validating is often so important.

If it is a large or largely remote deployment, you may want to consider creating a brief instructional video. This may seem to be a lot of effort for little results, but even a very basic video (delivered along with documentation) can go a long way towards increasing user understanding and enablement. Further, a video provides an opportunity for some commentary, while explaining steps. Many larger companies will hire out the video production for a truly professional experience that captures user attention. This is especially true for larger medical practices and task-centric companies such as insurance, where user buy in and enablement can be the difference between literally thousands of dollars saved per day with better user productivity.

We intend to publish in the future tools and document templates, and if there is enough interest, even conduct an online course for this process. Visit our Methodology Book's resources page at https://it.justdothis.net/bl for updates.

Chapter 17: Production Rollout

Updated plans are in place. Management and the whole team is ready. What is next?

Oddly enough, the Production Rollout stage should actually be the least amount of work. After all, at this point, you are essentially repeating validated processes. There are, however, a few areas in this phase where IT professionals often make mistakes, so this section will mostly be my advice to surviving a Production Rollout.

Adopt a Teacher Mindset. If you have followed this Methodology well, the Production Rollout step will not require a lot of "figuring things out." Instead, it is a fantastic opportunity to bring junior members into the process to learn the *why* and *how* of what is being done. If you are in management and reading this, I will tell you that for over a decade, I have noticed marked differences in reduced service escalations and increases in internal advancements when junior persons are allowed to take the time to observe a rollout. Even more valuable is when the junior person is able to do the work under the supervision of the senior engineer. Service Desk personnel wanting to become Subject Matter Experts (SMEs) are often eager to learn from you, so teach them!

This mindset is not limited to the IT teams. Teach a user WHY changes are occurring and what they can expect, and they will not only spread the word but they are much more likely to speak well of the change. Good management will take notice. When you get that promotion, email me!

Communicate in Advance. Particularly in major projects, advance communication with users and especially support staff is important. You should be informing what is happening as well as doing everything you can to explain WHY changes are occurring. What may seem obvious to you may not be to those that use the system daily. To help you, follow these suggestions:

- **Get Visual.** Whenever you can, include a comparison of how things look now and how they will look when the change is rolled out to the users. While small changes in things like icon appearance, position on the screen, and even font changes seem like something that is easy to adapt for us, the average end user will find this challenging. They are often coached to report things they see that are unexpected, because such things can mean the system has been compromised, so communicating the changes in advance and showing instead of telling goes a long way to reduce questions and increase confidence.

- **Get Verbal.** I recognize that the majority of IT professionals who will be reading this might be intimidated or feel they do not have the opportunity to "socialize" changes within their organization. In most cases I see, this is simply not true. Lunch and Learns, team meetings, and a host of other regular events often have the opportunity for IT staff to present an update. Yet, few take advantage. *Don't let when a user* *calls for help be the first time they talk to you.* One of my favorite TV shows is *The IT Crowd*. Every time I watch that show, I laugh, but I cringe at how awkward it is because it doesn't have to be that way. I'm not saying to try to be friends with everyone, but trust me when I tell you that if you can learn to communicate (yes, out loud) with people regularly and smoothly, you may find you have additional career options opening up to you. It is because management loves when someone can be effective, intelligent, and proactive. They also love it when someone like that thinks enough of the rest of the team to keep them informed.

- **Take it Slow.** Perhaps I should simply say to "take it as slow as you can" here. The primary goal of the Production Rollout step is to get the changes into production, but there is no such thing as a 100% risk-free Production Rollout—no matter how much testing is done. So, take the time to focus. Don't rush, even if you are currently behind. Hear me on this: Production Rollout is NOT the right step to make up time in your project.

- **Rollout in Groups, if Possible.** If you can isolate your rollout to smaller groups such as a department, floor, or other logical grouping, do it. By limiting the population getting the change at once, you have a better ability to back out, if needed, and a better ability to be present and in quality communication with that group.

- **Limit How Much Changes at Once.** I will admit that I'm an efficiency nut. If I can accomplish the same tasks in a shorter time frame by stacking them, I usually will not hesitate. However, the more you stack on at once, the more difficult Troubleshooting and Perception becomes. Remember that not all change is positive, from the user's perspective. If they are allowed to choose between learning two new things or one today, they'll choose one. When you have a lot of things going on all at once and something isn't right, you may find yourself rolling back ALL of the changes, causing people to have the poor perception of not having done enough testing. The sheer number of stories I have on this would fill another book! So, just don't do it.

Be Patient. This is a state of mind. When things seem easy or they are going well, you will have a tendency to want to please management by getting ahead. More likely, you are simply tired of the project and just want it to be done. So, during the action of taking it slow, patience is a mental exercise that you should adopt.

A secondary aspect of patience at this point in the process is being patient with those that are adapting to a new way of doing things. You will get questions you did not anticipate, and you will just as likely get questions you have already answered. Remember that while you've seen these changes for quite a while now, many users are seeing it for the first time and are overwhelmed with information. So, while communication ahead of changes is still important, the reality is that it is in our nature to forget that we've been forewarned.

Give Regular Updates. Remember the previous section about reporting Status? Although you may be tempted to become lax on this, or it may seem obvious, "assumption" is the most destructive force in modern IT. Be honest about your successes and failures and communicate any issues or problems early on.

Listen. I wish I didn't have to advise this, but when fatigue has set in and you just want the project to be over, it is a challenge to listen. We tend to want to believe all is well. Subconsciously, we may close ourselves off to getting further feedback because we can't accept that we need to make any further adjustments. My recommendation is to create a process instead of relying on intention. Have regular feedback-gathering sessions. Hear the hard things and respond. Your resume will thank you, in the end.

Chapter 18: Change Documentation

"Screenshot, or it didn't happen."

—*Common saying on internet discussions*

Just as with any of the other parts of the process, Change Documentation is still valuable. Even though the purpose of the *Change* phase is to have results that seem quantifiable on their own, this effect can lead to the incorrect conclusion that the results are adequate in and of themselves. As IT professionals, we must require some form of documentation to serve future needs.

We have defined WHY and WHAT should be configured. Now, we need to document HOW it was done.

What is needed here is to define first the documentation criteria. By now, you should have a good idea of what would best serve the company, but clarification is always warranted. Here are some basic criteria of a quality *Change* phase document:

Build Summary. This section is about what was configured (optionally, by whom). This can be as simple as a one-page summary memo or an executive summary used to lead into the larger document. On occasion, it is appropriate to attach a summary of Project Status Reports to this memo.

As-Built Document. Capturing HOW it was configured (or capturing a point in time) is the goal of this documentation. For this level of detail, I typically recommend an update to the design document. You are essentially telling a story here. If Design decisions were changed in the process, it may be prudent to document WHY.

Required Next Steps. Although the project may be complete by the time the documentation is being created, you may have identified other steps that must be taken. For example, if you have completed a project of load balancing websites in one datacenter but there is another datacenter for Disaster Recovery that was out of scope, you may want to note that updating that datacenter's configuration is an important next step. It may seem obvious to your current leadership, but you need to keep in mind that staff changes. It is better to be obvious.

Risks and Recommendations. It will often seem that if you have completed a project, all risks will have been taken care of. However, that is rarely the case. In fact, in my experience, I have typically uncovered more risks in the process of deploying software than I found in the *Understand* phase. Because we want to be protective of scope in the *Change* phase, it is important to call out risks you see so they can be addressed in another scope. Unfortunately, my most common recommendation is "Staff requires training."

An Updated Maintenance Plan. A negative feedback I very often hear in regards to documentation after a project is "completed" (built) is that the people receiving the documentation will have no idea how to maintain it.

Although this will be addressed in the next phase, and hopefully has some form of maintenance plan created in the *Plan* phase, a plan to maintain a built environment from Day One of going live should be present. In other words, as soon as production users are on the new configuration, you should be very clear as to how it will be maintained.

In-Process Documentation

Just as during the prior phases of *Understand* and *Plan*, I recommend a collaborative documentation process be used during the *Change* phase. The primary difference with the *Change* phase is that not all of what is documented will be kept for the final documentation delivery. Your goal with In-Process Documentation is to have:

Points of Reference of how systems were configured during each testing phase, for troubleshooting purposes as well as so you can refer back to how things were done and do them the same way again in another build.

Visibility into where you are in a process. In the event another team member needs to take over for you, having a running log of where you are in a process can be exceptionally valuable.

I have a few pointers to get you started on the way:

OneNote. As mentioned previously, OneNote has very good live collaborative features that allow you to take notes, document screenshots, and even put recordings right into a "note." The ability to be shared with multiple teams is very useful if your deployment is using a "follow the sun" method where a team will hand off to another team in another region or time zone, to keep progress moving forward during a 24-hour day period.

Screen Recording. Keeping a recording of what you have done can make referring back to the documentation a lot easier. This is especially true when you are working remotely, training, or not directly controlling the screen. When I am working remotely with a client using GoToMeeting (or any other electronic meeting software with the same capabilities), I will typically let them know I will be recording the session for my review purposes. Very often, they will ask for the recordings for training, which is fine. Several times, I have noted discrepancies in the final documentation which I can check after the fact by reviewing the videos. Drawbacks include the amount of space the videos can take, the length of time to find information in a large file, and the fact that this is hardly collaborative in nature. However, it is a great tool for going back and filling in screenshots. As well, I have often been asked questions after the fact and gone back to these recordings to find the answers.

Click Recording. A little-known tool built into Windows (since, at least, Windows 7 and Server 2008) is the Screen Recording tool. This tool was built for troubleshooting software development but I find it very useful during deployments because it captures each interaction upon clicking or hitting Enter. The best part is that the tool outputs a MHTML file that has tables of the interaction described (for example, where the user clicked) and a screenshot of the interaction.

A disadvantage is that this tool is limited to 99 screenshots and will not warn you when it has reached the max. The tool also must be launched as administrator to fully function properly and must be re-configured each time it is launched. However, the value of the output cannot be denied when it comes to both in-process and as-built documentation. More instructions and other alternative tools, when published, will be found on the Methodology Book's resource page at https://it.justdothis.net/bl. Make sure you register for our email updates to learn when these are published.

Now, it is up to you if you want to format your documentation in a standardized way, either for making a better process or for creating deliverables. Typically, teams will naturally gravitate towards ways of doing things that are helpful.

Build Summary Memo

Just as we note in the Documentation topic under the Executive Summary, your goal with a Build Summary Memo is to convey only the relevant information. List out the components configured. If it is relevant, you may want to include who configured each component. However, I recommend avoiding that particular information if you are allowed, because "finger-pointing" is never truly healthy. On the other hand, quickly noting who would be knowledgeable about the configuration beyond the documentation may be valuable. I would suggest if you are going include people's names, then do so in either a summary note or a footer for each that uses diffusing language, such as "Contact Team X for further information regarding this configuration, if needed." If you say something like "Team X was responsible for the build and configuration of this component," it can seem like you are pointing out flaws, even when you are not.

Other information you should include is a summary of the plan, if there were any variances, and when each component was completed (if relevant). Also, remember that this is your space to point out any high priority concerns that management will need to know, such as a lack of Disaster Recovery strategy.

Once again, keep it as simple as possible, using approachable language that is only as technical as it needs to be. That said, your leadership may want to see any of the following information in a summary table or other format:

- Virtual Machines used in Dev, Test, and Production
- Resources Consumed (storage, networking, etc.)
- Components or Virtual Machines yet to be built (or required at scale)
- Virtual Resources still required
- Risks and Recommendations

As-Built Document

The As-Built Document is the next expression of the Design document. This can be done as a standalone document or as an update to the Design document, if you are treating that document as a "living" document. Either way, the main goal of the As-Built Document is to capture a point in time of the configuration changes.

The As-Built Document should be approached in such a way that you could use the document as a reference to completely re-build the environment from scratch.

Useful criteria in this document are:

- Design Decisions, noted with how each was configured
- References to the location they can be found, if instructions are industry-standard ones.

- Descriptions of how the configuration was performed and why (if applicable), if instructions are not industry-standard ones.
- Tables with Diagrams and Screenshot Images (see the section on Diagrams and Images for more detail, which I will list, momentarily)
- Pre-Requisites
- Resources and Sizing (if appropriate)
- Scalability plan (if appropriate)
- Business Continuity and Disaster Recovery considerations or configurations
- Usage and Administration details (especially if they differ from industry standards)

If you are creating a new As-Built Document, approach it in the same way you did with the Design document—you will be telling the story of how the environment as configured meets the business's Success Criteria.

If you are updating an existing document, approach this with caution to retain credibility. Management will not appreciate an obviously re-touched document, especially if there are inaccuracies or if you used confusing language. Some points to remember are:

- **Change the language from "should" or "will" to past tense context.** Most design documentation uses a future tense. As-Built should always use the past tense. This is easy to miss, if you get into too much of a hurry.
- **Proofread carefully.** A temptation or mistake when re-using content is to let your eyes deceive you into thinking something was done that was not.

- **Do not neglect Quality Assurance.** Just as with the other documentation, you should always have the document reviewed by your peers or project leaders. They may have questions or items that require clarification.

Diagrams and Images

Pictures and diagrams are another powerful tool that can be used to illustrate your points. For example, if you are trying to describe an existing Active Directory structure, one of the best ways to do this is by taking a "screenshot" of the Active Directory Users and Computers display and placing that in your Control Layer section.

I would, once again, suggest bookmarking this book's resource page at

https://it.justdothis.net/bl, which will have more tools, tips and tricks.

I find it best to either send screenshots directly into the OneNote I am using, or send it to my clipboard so I can paste it into any application. In many cases, it is best to place screenshots into a table with descriptions, or to include a caption to each image. No image, diagram, or table should be without a description in your document. This is to prevent confusion with formatting. Word has powerful management for automatically labeling and changing numbers when you add captions after the fact; let the program do the work!

Why images? Images break up the monotony, for one thing. Let us be honest here, we are not writing thrilling fiction. These documents can be hard to get through in a single sitting. Images help the mind and the eye to look at your points in a different way, and provide further understanding. Diagrams (which we covered in Chapter 8) are also a powerful illustrative tool, so much so that the company may ask you only for a diagram as documentation, in some cases. Obviously, this is not your first go-to; you are here to add value, not take shortcuts. But this common request shows the power of the illustration. In Assessments, you may occasionally be asked to provide a high-level diagram of the current state. I typically like to include this diagram at the top of the document, right below the Executive Summary, when it is appropriate.

Caution: do not simply copy a grouping of screenshot images onto a page and believe that your intentions are obvious. A picture may be worth a thousand words, but your task is to point out what is not immediately obvious. Your audience likely sees what you are showing every day. They need to have text, describing a recommendation or observation, around the image. Use the tools to your advantage, not to your detriment.

Now that you have completed the *Understand*, *Plan*, and *Change* phases, it is time to move on to the final phase: *Maintain*.

Part 4: Maintain

Chapter 19: The *Maintain* Phase

Remember, our methodology has four phases:

1. *Understand*
2. *Plan*
3. *Change*
4. *Maintain*

We have covered the *Understand, Plan,* and the *Change* phases. Now that the changes have all been made, our focus shifts to the *Maintain* phase, where the goal is to ensure continued and proper operation. Although this may seem obvious, at times, the *Maintain* phase has similar planning and documentation as the other phases. The difference is that the *Maintain* phase targets a much longer time period. Naturally, in this phase, we will always be striving towards stability with low risks.

An important thing to remember with the *Maintain* phase is that it is always continuous. Even when we are in a new cycle of *Understand*, *Plan*, and *Change*, we must still *Maintain* what was previously in place. This is a primary reason why many companies will outsource the *Understand*, *Plan*, and *Change* phases to either contractors or consultants. Other companies may "insource" the tasks to a specialized team within the company. Some may even designate a Project Team or task force within those companies who are also tasked with maintaining operations.

However, without a good plan for which staffing and resources will be required to maintain operations, this can be either wasteful or detrimental. The IT team will need to determine the best course for maintaining during other *Understand*, *Plan*, and *Change* processes.

In order to *Maintain* properly, there are several key components for what needs to be done:

1. We must understand what Steady State should be.
2. We must establish baselines for performance, availability, and security.
3. A proactive monitoring system will be needed.
4. Knowledge of when the solution should be scaled should be monitored.
5. Report criteria should be established.
6. Business Continuity policies need to be established.
7. Data must be regularly backed up (data must be able to be recovered from a specific point in time).
8. Disaster Recovery plans will need to be created or updated.
9. Service Desk personnel will need to be trained (or established, if need be).
10. Service Level Agreements and a Service Desk solution need to be in place.
11. Points of escalation should be established for Administration, Engineering, and Architect positions.
12. Documentation such as runbooks should be established to train and guide staff.
13. Points of Procedure for when to re-engage with the *Understand*, *Plan*, and *Change* phases for new initiatives should be established.

Steady State

Now that your *Change* phase is complete and we have new processes for our users, it is important to catalog what normal operations should look like. This may seem simple but it can be much more difficult if you are not familiar with day-to-day operations. If that is the case, an additional Day in the Life study may be appropriate to discover what the "new normal" looks like to users.

From these metrics, we can establish Performance Baselines. For example, how much time should it take for a user to be able to log on? How many task items should the user be able to perform in a given time frame? What methods of access are allowed? What data security elements (compliance, etc.) must be monitored?

Related to this information is how the system itself performs when a Steady State has been reached. What we are targeting here is a Monitoring Baseline that will be used in our next topic, Monitoring. We will need to note how a system is reacting to normal operations over an established time period. I recommend this being at least two to three weeks, whenever possible, to accommodate for slow adoption and users getting used to the system. In many cases, as productivity goes up, so does system utilization. As users stop using workarounds and use your new system, resources will be consumed at higher rates. The point is to not make assumptions but to observe real-world expectations of the system.

Answering these questions and documenting them is what is referred to as establishing a Steady State. From this point, all tasks are to keep the status quo; to ensure operations stay as close to the intended state as possible. Of course, in any IT system, this is virtually impossible to ensure 100% of the time. Viruses, network communication losses, human error, equipment breakdowns, and any host of other problems can detract from this perfection.

Our goal will be to anticipate these problems and put systems in place to return us to a Steady State as rapidly as possible.

Other items to establish for your Monitoring Baseline would include Service Desk ticket averages and any known outage windows (planned or otherwise). We will explore more of how to track these metrics in the chapter on Reporting.

Monitoring

Monitoring systems are either processes, software, or a combination thereof with the goal of both alerting Administrators to problems and taking automatic steps to mitigate them, in some cases.

A good way to classify monitoring is by way of four basic categories: passive, manual, reactive, and proactive.

Passive monitoring is simply noting issues that occur, typically for reporting purposes. This kind of monitoring is quite rare but there are instances where it is required, such as closed systems that cannot be modified during operations—even to correct problems.

Manual monitoring is, unfortunately, the most common I have seen. Frankly, it is hard to call it true monitoring since it is about using more human labor because of a lack of automated systems. Manually scanning for issues that are occurring using non-automated tools is typical of this type of monitoring. It requires a good knowledge of a Steady State and what "known good" looks like when viewing performance metrics and other indicators. The primary issue with manual monitoring is that it is limited in scope to who is watching and what areas they are watching at the time. However, manual monitoring is often very appropriate for Service Desk individuals that require a limited scope and current point-of-time view.

Reactive monitoring is not always something that is targeted but is often the result of the aforementioned types of practices. Second to having no system at all (manual), reactive is very common. Reactive systems will scan for certain conditions (for example, when services are reporting as down) and alert administrators. Some of these systems will have the ability to restart services automatically and take other corrective actions, if appropriate. However, the thing to note about these systems is that they are correcting the problem or providing an alert to the problem *after* the fact; they are reacting to the problem instead of preventing it.

Proactive monitoring combines alerting with intelligence that will note deviations from established or observed baselines, in order to either automatically correct or alert administrators of problems *before* they occur. In the same manner, we can say that there is such a thing as a hybrid Proactive and Manual monitoring configuration when there is a combination of systems like automatic workload generation, overall system monitoring (manual and automatic), and even simply listening to trends in Service Desk calls, to determine when issues meriting an investigation exist. For example, an increase in storage latency on the Storage Area Network system may, by itself, be something that would not cause an alarm. However, extended amounts of latency in certain systems can lead to cascading issues with slowdowns and even service failures. If the latency is identified and dealt with early in the process, it can prevent additional issues from occurring. Because we have taken steps to mitigate before the actual problems occur, we call it Proactive.

Artificial Intelligence (AI) is new enough (as of this writing) that I hesitate to say much more than how exciting it could be, especially in a security context. Noting changes in behavior rather than exceeded thresholds has its share of potential frustrations and benefits. I do not see a downside when it is used for small contextual lockdowns and smart alerts. Of course, people think of Hollywood portrayals of AI which slows management adoption.

Realistically, the potential for moving baselines, smart recommendations, and suggested workflow enhancements has the potential of shifting IT Administration to the order of what Eli Whitney did for cotton.

What I will say is to not confuse AI adaptive solutions with rule-based proactive systems… for now.

The best monitoring solutions also reduce the amount of staff effort required. At the very least, they make the Support Desk staff able to more rapidly route and resolve issues, thereby reducing the impact of disruptions.

As you can see, I have not yet mentioned reporting. We will be touching on that in another section, soon. However, many of the best monitoring tools also have reporting and tracking capabilities.

You might be wondering, *Why isn't everyone using a hybrid Proactive Monitoring system with good reporting capabilities?* While complicated, the answer is generally the cost. The best systems do not come cheap. So, how do you determine the value, especially when you need to convince those in control of the budget? As with many such decisions, taking the approach of starting with a business justification typically works best.

The first thing to do is to start to determine the business impact of outages. Both large and small disruptions should be analyzed, in this regard.

How much does it cost the business if users cannot work for a certain period of time? Is there an impact to long-term business, such as customers not coming back or going to a competitor instead? Are there Service Level Agreements or uptime requirements given to customers that would have additional payouts involved, if uptime requirements are not met?

Second, determine how often are outages happening, and how long are they occurring. I recommend trying to estimate this time frame over the course of a year, to determine the overall cost of outages in a budget year. Again, we are talking in management estimates, not actual data.

Third, determine how much staff time is lost, either in manual or reactive monitoring, and if this is impacting overall costs.

Finally, determine the cost of the solution(s) to implement and determine if this cost is less than 50% of the overall outage costs. Why 50%? Because no monitoring system is able to prevent all outages, but a good system should be able to offset at least 30% of outages and reduce staffing costs by another 20%. I will say that some of the market leaders often see performance far above this; in fact, I have seen the implementation of certain solutions that reduce time to resolution of issues and true environment-wide proactivity reduce costs more than 300%. Better yet, I anticipate AI-based solutions or hybrids, which will easily exceed this, freeing staff to learn and be better utilized for more advanced and rewarding tasks.

Now you know why I was excited about this section!

Chapter 20: Scalability and Performance

IT Architects are usually tasked with a seemingly impossible task: determining the hardware and software requirements for a user environment that is under near constant change. Determining if the current or future physical assets are adequate for the changing demands of a business is a top priority for management.

Likewise, Architects may also ask the system Engineers to monitor the overall performance of the system on a regular basis and generate reports. The Architect then takes these reports, compares with previous Key Performance Indicators (KPIs), and determines if there is a need to re-enter the *Understand* phase.

The reality is that someone needs to be proactively thinking about these things or the company is at risk of falling behind.

The large problem is that when it comes to words like "performance," this is sometimes subjective; this is especially true when custom applications are in place that the team has difficulty establishing baseline performance expectations around.

Nonetheless, this is the task at hand and it is very important to put together the right set of tools to address the need for meaningful direction.

An additional note here—specifically if you are in the Architect role or similar—when reporting to management is needed is to remember that pictures can communicate what words fail to, so gathering and presenting data in front of the C-suite will make you exceptionally valuable to them. To have something to present, however, you may need to upgrade your skillsets in programs like Excel and PowerPoint, to draw the meaning out of the data more potently.

Load Testing

As the name implies, Load Testing is the process of purposely introducing a load onto a system, to determine a response. Load Testing can have several goals, but the most common are determining baselines for expected performance, determining the expected scalability of the solution, and my personal favorite: proactive application monitoring. There are a variety of tools in the marketplace that address these needs such as Shopping Carts to SaaS applications and VDI environments.

In all cases we will be discussing, there is a synthetic workload introduced by a technology that simulates working conditions, then patterns and measures the results by using a *launcher* or *workload generator*. The good news here is that you should already have a testing plan you can use for getting started in writing your *testing script*.

The script is a set of instructions that are given on a consistent basis, each time the test is run. Depending on the technology, you may have some built-in scripts to do some testing but for true monitoring, you will want to either modify those or use the tools to create your own test pattern.

Typically, the system will integrate with your infrastructure systems as well, to generate reports that correlate performance observations with infrastructure conditions.

Common metrics gathered include:

- Application Load Speed
- Site Responsiveness (time to respond to clicks or input)
- Logon Time
- CPU Utilization
- Memory Usage
- Disk Input/Output (I/O)
- Network Bandwidth Consumed

Performance Baselines

After a change is complete, but with few or no users on a system, baselines are created by simulating a variety of workloads over a relatively short period of time. The goal is to determine what to expect of the system under normal conditions, depending on your associated business goals.

For example, if a business goal is to achieve the maximum number of users on the fewest servers possible, the system would be asked to simulate as many user workloads as possible before performance suffers (often referred to as UserMax). It should be noted that an important criteria here is to determine which metrics are important. For example, if the response time of a webpage loading is important, then that should be used to determine your UserMax. In some cases, mouse or keyboard response is the most important metric. Others may be logon times.

Another possible goal is to determine the breaking point of a system; typically to determine the Service Level Agreement that will be appropriate. In these kinds of stress tests, the system is subjected to a relentless amount of users, until it simply fails and the last metrics recorded are noted. This may be referred to as a LoadMax (which is often incorrectly referred to as MaxQ). Some similar tests may include establishing at what point the UserMax is affected by other systems. For example, an engineer may introduce a synthetic IO workload to the storage system, to determine at what point it will negatively impact the UserMax or LoadMax—common in database testing for high-volume web apps.

You can likely think of several other possible criteria here, but the most important thing is that you are able to establish real-world examples of what the system is capable of, so that you can properly monitor, without guessing. Especially in the realm of UserMax, knowing in advance when users will likely start to complain is extremely useful.

Scalability Testing

Similar to establishing a performance baseline is establishing how much a system can take and still be productive, and establishing the impact of the loss of a resource to the baselines. This is often done by establishing how much service volume a single server can handle, in order to predict the resources required for a specified workload. For example, the goal may be to establish how many physical servers need to be purchased per every 1000 VDI users that are added. While the Conceptual Design (or deployment, depending on when the testing is conducted) may have estimated five physical host servers to handle this task, for example, you may find by simulating your actual workloads on a single server that you can only achieve 180 users per host. This would then mean you need six servers per thousand users (and adding one or two into the total count for redundancy).

The key here with Scalability testing is more financial and capacity-planning oriented. It should be noted that you are always going to estimate UP, not down, because— unlike establishing UserMax/LoadMax baselines—you are not trying to predict overall performance but are estimating *when* it will be time to purchase more resources, based on workloads.

There may be other cases where you need to determine when it is appropriate to shut down resources to conserve power or spend less. In Cloud scenarios, such as Microsoft's Azure, for example, you are charged for the time a system is running. The ability to correlate application usage metrics with the servers required in a scenario (with varying but predictable workloads) can greatly increase ROI. Automatically shutting off workloads during low use times saves money. Even though this is probably less common, you may find value in doing that, to be able to answer questions in advance for your management regarding the classic "How much will it cost to add another 3000 users?" question.

Proactive Application Monitoring

The testing I have mentioned up to this point is typically done after a change is made, then left to the occasional follow-up as production continues. However, the investment in time and resources for load testing can have one other very important use in monitoring production applications. How this works is you will set up a series of lightweight tests (simulating a single user, typically) in various simulated or actual scenarios to observe response times and other metrics. These metrics are compared against baselines and if they vary, you are notified in advance. Proactive Application Monitoring would identify, in advance, when trouble is occurring and notify you before the Service Desk calls start to come in.

For example, let's say you wanted to monitor an ERP (Enterprise Resource Planning) system that is accessed worldwide, but you've come to realize that by the time a service has failed and you are notified about it (likely by the Service Desk), potentially hundreds of people are already not working.

However, in a Proactive Application Monitoring scenario, you may have launchers in each office building. So, if *one* launcher starts to report slowdowns, you are more likely to suspect something with their network, instead of the system as a whole. But if *all* of the launchers start reporting slowdowns or not being able to log on, you know right away that you need to start your troubleshooting in the datacenter.

I'm willing to bet after reading that if you don't already have this set up at your practice, you are likely wondering how you never knew that was possible. Well, I just described a scenario that happened in real life. The on-call Engineer was notified of the issue at home and took care of the problem, notifying the Service Desk in advance. The Service Desk was able to inform users that the problem was being worked on but never had to actually do so. The Engineer later told me that the same thing had happened the previous year and they were down for more than a day because of the cascading nature of the problem. In other words, they paid for their investment in the Load Testing software some 5,000 times over.

The combinations are nearly endless here. For example, if you have a region experiencing heavier loads than another, based on your Scalability Testing, you know that you can simply spin up additional workloads to help. But without active, automatic, and persistent testing that happens without your input, you may not have a good idea of whether or not the extra resources you have spun up are having a positive impact.

Proactive is, in my opinion, the single most important way Information Technology should be done. So, for practical monitoring, having this sort of information at your disposal is crucial for getting there.

Chapter 21: Reporting

"You won't know, unless you ask."

—Unknown

Reporting is the process of regularly gathering and documenting performance, usage, alerts, and even Service Desk trends that have occurred in your environment over a certain length of time. How detailed the reports are and how often you report is typically up to management to decide what is valuable. Often, these reports are driven by Key Performance Indicators (KPIs) that are based on a balance between management's expectations and the Performance Baselines you determined when the Steady State was established.

There are a few reasons you should regularly report, even if management has not asked you to. Primarily, you should be aware of trends (positive or negative) in your environment so you can adapt or, hopefully, be free to work on other tasks. If there are negative trends, especially ones you can detect and be proactive with, this can be a very positive growth for your career path (trust me!). In addition, if you can help reduce Service Desk calls, your reputation will be stellar. Although I don't want you to be prideful, if you have taken the time to invest in upgrading your skillsets and are thinking proactively when others are not, you need to be given the proper credit. Generating reports are the perfect way to do this, because you can simply state the facts when you have success. This same data can even point out when you need additional staff, using facts instead of opinions.

In 2019, I spoke to a CIO (Chief Information Officer) for a large healthcare organization who told me that it didn't matter to them what opinions their staff or vendors had, and that without regular reports (data), their budget was not going to move. To many CIOs, the budget is actually a primary consideration, not the technology in use. If you want to see movement, a new project, get a raise, or get help, then you need to be able to prove it.

So, what should reports contain and how do you generate them? Rather than start guessing, I will point you back to the Business Needs Assessment at the beginning of the *Understand* phase. (Hopefully, by now, you can see why you can't skip bits of this Methodology!) Your reports should center on delivering to those needs.

In addition, ask either management, Architects, or even yourself, *What indicators of success and trouble do I need to see? What technical areas can positively or negatively impact that success?* If you are not monitoring these areas already, please see the previous topic on establishing Monitoring (in Chapter 19)!

For example, if your Business Need was to ensure users could access a key internal application from their iPads, we have several key areas to monitor from our built solution. We know we need to measure components like:

- inbound Internet connection uptime

- Application Gateway uptime
- application uptime
- logon times
- application performance (response metrics from load testing)
- Service Desk incidents (per week/month/quarter)

Along with the possibility of a user survey, you have metrics that you can put into a table and compare against baselines. This is a good basis for a report indicating if the application requires attention or if all is well. It can also indicate if the solution may need to be scaled.

Chapter 22: Business Continuity

Very often misunderstood, the need for Business Continuity goes far beyond a simple "backup" system. Far too often, I have seen companies falsely assume, for example, that simply putting their environment "in the Cloud" will provide the ongoing resiliency that is required to keep the business running.

Another common confusion is that Business Continuity and Disaster Recovery are the same thing. They most certainly are not, even though they have similar themes.

A proper Business Continuity strategy needs to address how normal operations will be maintained and risks will be avoided. In fact, whether it is written down or not, this Methodology, in and of itself, is part of a Business Continuity strategy!

By defining how we avoid risks to production, such as a phased approach and constant follow-ups, we are keeping the business running under a normal state for as long as possible.

Although I have seen a great number of companies over-complicate or under-complicate this process, the

Understand, *Plan*, *Change*, and *Maintain* strategy—applied to all aspects of operations—should be expressed in a policy the company has. This, by the way, is a common failing of those attempting to adopt an "Agile" or DevOps strategy incorrectly.

Keep in mind that how to avoid outages is one thing, but a general attitude of avoiding any disruption to the user's daily operations is what a true Business Continuity strategy or policy should be focused upon. Rapid changes should never come at the cost of user productivity, even if they weren't intended. This also, in my opinion, is where the DevOps model of Minimum Viable Product falls apart. From what I can see, user satisfaction has been dropping in products that adopt this model. A services-oriented practice should not follow this and settle for the bare minimum service. That said, it is a great way to make sure you will be replaced, if that is your intention.

But that's only part of the big problem facing IT.

A 2013 study by Quarum estimated that failures were caused:

- 55% by hardware failure
- 22% by human error
- 18% by software failure
- 5% by natural disasters

A 2016 study by Zetta found both impacts and causes point to several points of concern:

- 54% of companies reported that they had a downtime event lasting more than 8 hours (which I consider Business Continuity, not Disaster Recovery, by the way)

- 67% of respondents said that the daily impact of an outage would exceed $20,000 USD (every DAY of downtime!)
- 19% would lose between $100,000 and $500,000 USD, and 8% would lose over $500,000 USD per day of downtime (!)

The causes of downtime events was also very interesting (note: respondents were allowed to respond to multiple causes):

- 75% reported power outages as causes
- 35% were due to human error
- 34% were due to malware or security attacks
- 20% were due to natural disasters
- 11% were due to on-site disasters

Further complicating the matter, a 2017 article by StoneFly found that many companies they surveyed had put the technology ahead of the actual business need. They looked at business disruptions as a whole, as they relate to a BCDR (Business Continuity and Disaster Recovery) strategy. Their results found that:

- 44% of disruptions were caused by hardware failure
- 32% were due to human error
- 14% were from software or firmware errors
- 7% were from malware or other security breaches
- 3% were due to natural disasters

For the sources of these studies, visit https://it.justdothis.net/bl#7.

Are you seeing any patterns? While things may be getting better, this resonates with my observations as well. It amazes me that any company would consider only lower priority items such as preparing for Natural Disasters (which we will talk about in Chapter 24: Disaster Recovery) for their BCDR plans.

You need to consider the following aspects for an effective Business Continuity strategy:

Redundancy. Every production system should have layers of redundancy for both load balancing and for ensuring availability to users. If one component goes down for any reason, another should take the workload, automatically. Another way to say this is to ensure there are no single points of failure in compute, network, storage, and even in staff.

High Availability. Related to redundancy, High Availability (HA) is a term applied to how workloads will either automatically restart if down, or automatically re-direct workloads to other systems, to accommodate unscheduled outages. Given how often software errors occur, this is very important, as is proactive monitoring. An element that should be mentioned here as well is load balancing. HA systems that are not load balanced often face unexpected outages, if they are not redundant and load balanced, because workloads typically don't automatically shift.

Workload Separation. Especially in virtual environments, redundant components should not be housed on the same hardware. Remember, it was found that 44% of outages were from hardware failure. If the redundant components are on the same hardware and it fails… what was the point? Many organizations will go so far as to put rules in place, separating components into separate floors, datacenters, racks within the datacenter, or other ways of ensuring the platform is as isolated as possible, to avoid this most common type of outage.

Resiliency and Stability. Although it should go without saying, a production system should be running on reliable equipment. However, other components above the hardware layer should also be focused on being "stable." In other words, experimental or otherwise unproven software, hardware, or operating systems will rarely be appropriate because they are not fully known to be stable. You should not expose your users to this risk, and have policies that support this by requiring warrantied systems that have been validated, either by testing or industry confidence.

Uninterruptable Power. While the most economical option, reliance on public utility for power in a datacenter that requires a heavy amount of it is not always wise. Power disruptions in the world of servers that should be up 24 hours per day can be absolutely devastating. Even a very brief disruption of less than a single second can cause massive disruption and even damage to servers.

Since the late 1990s, most companies interrupt the public utility with a continuous power supply that is able to instantly switch to a battery until the power returns to normal. This prevents servers from seeing a power outage from either a loss or drop in power levels. However, batteries cannot run for very long, so servers will either need to be shut down gracefully in the event of an extended outage or expensive generator systems will need to be run. Since power is recovered in under a day, in the majority of cases, I consider a power failure to be a Business Continuity preparation because the plan typically does not involve shifting workloads to a new location. That being said, many companies that rely on real-time data may opt to move workloads freely between multiple centers in the event of a power outage lasting more than 20-30 minutes.

Recoverability. The most common use in backup software is actually not recovering entire systems (Disaster Recovery) but in recovering individual files or snapshots. Your system should properly accommodate this by setting rules about how data will be backed up and when. It should also be noted with the prevalence of human error, malware, and software errors that a good data recovery plan requires special attention.

Based on your Service Level Agreements (which we will discuss in depth, momentarily), you will need to determine how often to back up data but also—just as importantly—how fast you can get data back in place and who will be responsible and trained to do so at a moment's notice.

Change Control Procedures. Our Professional Services Methodology was developed to work with companies wanting to address the second-largest (and in some cases, the largest) concern: human error. A quality Change Control should involve *Understand*, *Plan*, and *Change* procedures that require documentation, peer review, and most importantly, testing, prior to being put into production. I believe that if more companies properly adopted and adhered to their Change Control procedures (shockingly, few do), we could see that 32% down to more like 10%. The key is using the Methodology and not skipping steps to save time.

Service Level Agreements. Crucial for those providing a service to other companies, a Service Level Agreement (SLA) defines key aspects of Business Continuity. For example, an SLA may include:

- **Defining Your Overall Downtime Tolerance.** This important metric is typically defined over the course of a year. So, when you hear or see people refer to "five nines," they are referring to a system being up and available 99.999% of a year. To save you the math, that is only 5 minutes and 16 seconds of downtime in a whole year. "Three nines" (99.9%) would be almost 9 hours. If your SLAs are defined in a different time frame, the expectations are drastically different. For example, a daily SLA of 99.9% means the system can be down for about a minute each day. Regardless, if the business leaders have this expectation, it is important to take steps or adjust your plan to match.
- **Time to Recover Data.** For example, how long can users be without having the data recovered?
- **Data Backup Information.** Here, you can include the points in time that data is backed up or how often data must be backed up. (More on this in the next chapter on Backups.)
- **Service/Help Desk Response Times.** We will discuss more about this in the Escalation Procedures topic in Chapter 25, but SLAs for a Service Desk generally outline how quickly a trouble ticket is responded to and possibly defines an expected time for resolution. Guiding this SLA is often crucial for determining the amount of staff required in a Service Desk.

Performance Benchmarks. In the Monitoring section, we discuss more about how to determine expected baselines, but many systems in key sectors such as finance and consumer-facing websites will require a certain response time to be maintained. This requires a further series of validation against moving baselines. This process is typically referred to as "benchmarking" (testing the current configurations against the last or reference baseline). We do this to determine the impacts (positive or negative) of changes that have been made. For example, if a series of program updates are scheduled, benchmarks are run against the new configuration to determine what to expect from the updates scheduled. While not every organization requires this level of effort, the ones that do typically have fewer support calls about experience changes because they are either able to communicate the expectations in advance or adapt to the new requirements to match prior baselines. The reason we include this within our SLAs is because we need justifications for asking for additional hardware to support changes being made, especially if there are actual performance and response time metrics that must be maintained. There are a few key tools to know about for this kind of testing. Find out more about them by visiting https://it.justdothis.net/bl#8.

Chapter 23: Backups

Data backups can come in many forms, but the underlying principle is that the bits or bytes of information that exist in a Steady State are written to a secondary form of storage that can be accessed and recovered within a specified manner or time frame (as per the SLA for data recovery, see the prior chapter).

Remember that we observed earlier that human error is the most common cause for recovery needs. In numerous cases, over the last 20+ years, I have seen this expressed as "I accidentally deleted this file/folder" or a similar accusatory statement. No matter what the cause, being able to recover individual files or restore systems to a point in time before a change was made is a vital part of an IT strategy.

Backup media has changed a great deal over the past 50 years, including the increasingly popular Cloud options. You must be able to understand the implications of each technology you are using and generate backup procedures for your staff to use.

Your design should have included the answers to very important questions: What items should be backed up, how frequently, and what is the Service Level Agreement on recovery of data? If your design did not include that information, then surely you know by now that it is time for another cycle of *Understand*, *Plan*, and *Change*.

Once the business has defined your SLAs overall, you may want to map out recovery SLAs for each application. This information will be used to determine the right fit for the backup system used for specific targets. Much like Disaster Recovery, these SLAs should include key metrics of:

- **Recovery Point Objective (RPO).** In this case, this metric essentially defines how often data is required to be backed up. In many cases, data will be backed up more frequently than is required, which is good.
- **Recovery Time Objective (RTO).** In this case, this metric defines how much time the administrators will have to recover the data before the application is at risk. This will often also define the number of staff assigned to backup recovery duties.
- **Retention Policy.** This time frame, similar to RPO, defines how long data must be kept available for recovery within the RTO. For example, data from a point of time three months ago may be called from and still have an RTO of one hour. IT needs to be able to know the expectations and set a policy for both on-site and off-site RTOs.
- **Items to be Backed Up.** While some companies may simply opt to back up everything on a machine, it is still a good practice to define what data requires a backup and that which does not actually require backups. For example, in many cases, the underlying Operating System can be rapidly recovered or reinstalled.
- **Backup Targets.** This could include file paths, machines, or other targets. It defines where the data

being backed up currently exists. For example, redirected Windows folders may live on the SAN, while the databases are on iSCSI LUNs. Or at times, physical servers are still in use and will need a backup agent installed.

- **Type of Backups.** Some backups are simply flat files written to a storage system. Others are Virtual Machine snapshots. Databases may require a more complex agent-based backup that sets a specific transaction log point in time and backs up data at that point, resetting the transaction logs after the full recovery point has been ensured.

- **Backup Cycle.** Depending on the backup technology used, a cycle of backing up a Full State followed only by the files that have changed is quite common. This is usually referred to as a Full and either differential or incremental backup. Differential is all data since a Full date. Incremental is all data changed since the last full, differential, or incremental backup was taken. Because these structures can be complex, the risk of backup loss or corruption is higher. However, especially in the case of incremental backups, the network and storage consumed during the backup is significantly less.

- **Off-site Retention Policy or Cycle Frequency.** While often confused with Disaster Recovery because your off-site backup cycle may include or consist of shipping data (physically or electronically) to your Disaster Recovery environment, this is not always the case.

  o Many companies will, on occasion, ship large tape or hard drive collections to an off-site storage with a day or less of time to retrieve the media in the event a larger recovery event or point in time past the retention date is required. This is essentially its own RPO/RTO that needs to be defined. This is typically how often full backups are swapped.

o Additionally, if you use a Cloud strategy, this still applies. How long will individual data sets be kept? How often will the expensive data transfers to the Cloud take place?

- **Compliance.** Several governmental policies may control how long data must be retained as well as the sovereignty of the data (when data must remain within the borders of a country).
- **Archival Retention Policy.** To meet certain compliance needs but not have active data, you may be asked to set up a data archive. These typically have a "best effort" RTO and an RPO, as far back as the compliance requires (7 years not being uncommon).

Obviously the names may vary here, but I typically recommend constructing a spreadsheet, table, or database for these requirements and reviewing them any time a new backup system is proposed or new requirements are set.

Define all these criteria for each application, and then sort by the requirements of each column, to determine the overall highest needs. You will then need to determine the technology, cost, and staff required to maintain that, and document instructions on how to recover files. This documentation may be subject to certain regulations—this is true of HIPAA (Health Insurance Portability and Accountability Act), in the United States, for example, which requires those giving patient care to have a documented recovery procedure in the event that a patient asks for data.

Chapter 24: Disaster Recovery

"You will not rise to the occasion.

You will only rise to your level of preparedness."

—D.J. Eshelman

Disaster Recovery is the procedure of moving Compute, Storage, and Networking operations to a completely different datacenter than was previously used.

Not many people enjoy talking much about scenarios where an entire datacenter is disrupted permanently, yet the loss of business suffered when a massive disruption occurs makes the topic of Disaster Recovery one that every business should have. The most common disastrous scenarios thought of are weather related, even though weather is rarely the cause.

Recall the previous studies we noted at https://it.justdothis.net/bl#7.

The 2016 Study by Zetta found:

- 54% of companies reported that they had a downtime event lasting more than 8 hours

- 67% of respondents said that the daily impact of an outage would exceed $20,000 USD (per DAY of downtime)
- 19% would lose between $100,000 and $500,000 USD, and 8% would lose over $500,000 USD PER DAY of downtime

Also, remember the patterns of typical outages, in order of most to least common:

1. Power outages
2. Human error
3. Malware or security attack
4. Natural disaster
5. On-site disaster

This means that even if it is something you hope will never happen (and by all rights may *never* happen), most companies that do not have a viable Disaster Recovery strategy could find themselves closing down… all because they did not invest in the extra "insurance" of Disaster Recovery.

Massively disruptive events happen more often to businesses that are trying to save money by hosting their own solutions or not testing the solutions they have frequently enough, depending instead on staff knowledge. When events occur, they find that they are unprepared, possibly with staff who do not know how to recover, simply because they don't have a recovery strategy or it has not been documented properly. This is tragic, but it happens.

Importantly, we must remember that **Disaster Recovery is NOT a *technology* decision but a *business***

decision. As such, the business must drive what is acceptable and not acceptable for Service Level Agreements. So, in preparing such a strategy, to accommodate these SLAs, the common question to Business Leadership must first be "What is our priority?" and *then* "What is our budget?" However, and I cannot stress this enough: BOTH questions must be answered, or you will not have a solution.

In Disaster Recovery, there are two key metrics that may vary per application but must undergo the same process of *Understand*, *Plan*, and *Change*, put into practice and documented:

The Recovery Point Objective (RPO). In Disaster Recovery, the RPO defines the point in time from which your environment is recovered. Note that this is not a frequency of backup per se, but rather when the said backup is available in the Recovery Datacenter or Cloud. So, if we are only replicating data from our weekly midnight backups, even if we have more frequent backups, our RPO would be one week. Naturally, we need to ask the business leadership if losing a week of data is acceptable. If is not, we need to define a different RPO and assign a cost to the increase in replication time frames.

The good news is that bandwidth and backup/snapshot technology has greatly evolved to be much more practical for tighter RPOs than ever before. However, RPOs less than a few hours become much more difficult and expensive. The need for real-time database transactions to stay relevant may mean your point in time is less than a few minutes. So, your costs will be very high but your back-end technology must support it, in these scenarios.

For these, we would have closer to an Active-Active (Hot+Hot) datacenter configuration, which we will discuss in more detail shortly.

The Recovery Time Objective (RTO). The RTO defines how long IT has to recover operations to the Recovery Datacenter. In this, leadership is defining the length of time the business can be without active systems. For example, if it is determined that the business can survive for three days without IT systems, your RTO would be three days. Again, this comes down to a "cost versus risk" scenario. Being ready for a rapid RTO is quite expensive, in most cases.

- "Cold" datacenters are those without active workloads (in most cases, not even powered on), which must be brought up or recovered from backups. With proper preparation, this can be extremely cost-effective and have a relatively short RTO. These are sometimes referred to as "passive" datacenters, as they can be configured to serve almost any need, given the correct time frame. This is not uncommon in a Cloud recovery scenario where actively running machines are more costly.
- "Warm" datacenters are those that have slightly out-of-date servers or data in place that must simply be updated with the latest RPO. Most warm datacenter configurations operate on minimal required power (one of each component, for example) but keep active copies of technologies, such as an Active Directory, at all times. This is the most common configuration I see in American enterprises because it balances a relatively low RTO with acceptable costs. Some are even able to run a blended Active-Active scenario

with certain applications, relying on the link between datacenters to function as a Hot+Warm configuration.

- "Hot" datacenters are fully active and serving workloads. Here, in many cases, we would be serving a near-instant RPO and have essentially an RTO of less than a few seconds.

Running in an Active-Active (Hot+Hot) scenario very nearly negates a Disaster Recovery Plan, in many cases, although most companies that can afford this kind of configuration do not take chances and still have a plan for recovery. Active-Active configurations, as previously noted, do not have a back-end dependency and can run workloads virtually anywhere (or in any Cloud). Others have dependencies that must be taken into account from data distance and data cost standpoints. Most real-time systems like this are essentially writing data to two locations at any given time (sometimes more) to ensure absolute data resiliency. Technologies such as Global Server Load Balancing (GSLB) are used to intelligently direct workloads to the appropriate datacenter. Interestingly enough, if an Active-Active configuration is properly load balanced (with about an acceptable RPO), it can represent a significantly better ROI (Return on Investment) than Hot+Warm or Hot+Cold configurations, because both of those scenarios involve money being spent that does not have immediate results. However, an important caveat to running an Active-Active scenario is that you must design for full-scale operations at both datacenters to be active at nearly all times, unless you can declare a brief RTO to prep workloads before they enter the active datacenter, in the event the other is no longer available. This can happen very frequently with inbound Internet or MPLS failures.

The problem, in most cases, is that your company will not want to spend what is required to bring workloads to a truly low RTO/RPO state. Therefore, you must find the proper balance, then design for and implement that solution. Documentation is important for a unique reason, as it is not something that is actively accessed on a daily basis and therefore, will not be "front of mind."

Disaster Recovery Plans must be updated on a regular basis, especially as technology evolves and needs change. If your company is publicly traded, you may be required to declare how often you review these strategies. I typically recommend reviewing them at least once per year. However, I believe companies should not adjust the strategy too frequently, and only do so after testing. Quarterly adjustments to the overall strategy may be impractical as well as take a lot of time and expensive effort. Related to this, if your system is not tested regularly, you are at risk; yet testing too frequently can be disruptive to operations in and of itself. The key is to be properly prepared, in a practical way.

Chapter 25: Data Security and Integrity

"My voice is my passport. Verify me?"

—Werner Brandes, fictional character in the movie *Sneakers*

Data Security and Integrity are certainly very much in focus, in the Information Technology world. As more and more companies experience data breaches, it has become very clear to me that simply a good initial design is no longer adequate. In addition, the time of mere virus protection has passed, as threats have become far more malicious and adaptive than IT has. This means that the need to properly maintain systems with a focus on Data Integrity cannot be done passively. You must be active and vigilant.

Much as we described earlier how Canadian Police are trained to spot counterfeits, not by looking at the different ways that fake currency can be made but in studying the way it should be, Data Security starts by knowing first what a Steady State is. In this regard, merely trying to understand all the different ways you could be compromised does not make for good security.

You might say that good Data Security is a Professional Services Methodology all to itself—regularly assure that you *Understand* a system state, *Plan* (design) a solution to deal with threats, and enact a *Change* to *Maintain* security. This process may happen in milliseconds or over the course of a comprehensive scan. Chapters could be written on this topic and I am not fully qualified to write them. (However, I will say that subscribers to my email list get occasional notifications when my friends and trusted resources make content available! Oh, the joys of professional networking! You may subscribe to that list and also access the bonuses for this book at https://it.justdothis.net/mybonus). What I simply will state here is that your company (read: you) should have a *Plan* and a system to ensure that data is secure and that it remains what it is supposed to be. In other words, not all file alterations are caused by malware or malicious intent. You should be proactive in regards to identifying file corruption as well as infection, so that you can rapidly respond to bring the system back to a Steady State.

Here is a basic list of threat types to be monitoring for, as a baseline:

- Viruses
- Malware
- Trojans
- Spyware
- Backdoors
- Keyloggers
- Exploits
- Botnets
- Phishing
- Distributed Denial of Service Attack (DDoS)

- Spam

If you cannot ensure protection against these threats, 24/7, then it is time to have a fresh *Understand*, *Plan*, and *Change* cycle!

However, the unfortunate reality is that not only will that list become antiquated before I have finished editing this book, but these may not be the biggest concerns that you will face. The business landscape is becoming increasingly aggressive regarding outright sabotage and theft of information. Employees and trusted contractors with access to secured systems get angry and seek to hurt the business. They may be paid to steal information. In a shocking number of cases (which are increasing!), individuals able to pass a security background screening are actively working to set up information theft that may not occur until months or years after they have left the company. This means that you must be vigilant about outside threats (as previously described) and you must also protect the system from within, which may mean taking steps to limit access to key systems, or—in my favorite scenario—outright restricting any datacenter communication to only specific server-based applications or desktops that are tightly controlled and reset their operating system back to a "gold" image at each reboot.

The "trust but verify" notion is not always appropriate in IT because of the final threat we will talk about, which is when users are not acting maliciously but their credentials or systems are being used in that way. Taking steps in ensuring logins happen with the actual user's intent is important.

Multi-factor login steps should be part of the design, for sensitive systems.

- Authority must be derived. For example, usernames establishes identity and passwords establishes authority. A PIN may establish a secret, known in a specific context, such as a phone or computer having an internally stored PIN that is not given to the server.
- Possession is next. A rotating token or passcode sent to a mobile device would establish possession, in many cases.
- Finally, you may want to also ensure inherent living conditions (life), by using biometrics, voice print identification, or retina scans. What was once science fiction or seen only as securing the most crucial government secrets in movies are often used by small businesses nowadays.

For example, most physicians in the United States are required to give a password, a PIN, and a fingerprint scan, in order to prescribe drugs electronically.

But all of this prevention is worthless if it is not maintained and used properly. That should be our goal—along with regular security assessments, penetration testing, and other proactive measures.

Chapter 26: Operational Support Plan

All this talk of keeping the system running at the best it can is wonderful… but who is going to do it? Part of your design in the *Plan* phase should have included parts of this in the Operations Layer, but in many cases, a separate Operational Support Plan (OSP) is required. Regardless, now that we have defined the *what* of the *Maintain* phase, the Operational Support Plan should include the *who* and the *how*. For many years, I referred to this as an Operational Support Design, but I felt this did not always cover the final element of *when*. Every *Plan* should have a time frame for events, and the Operational Support Plan (OSP) is no exception.

We have already discussed Service Level Agreements (SLAs). The Operational Support Plan will take things several steps further by defining the staff, methods, and technologies used. It will also define how often reports are generated and Key Performance Indicators, if required. Additionally, a quality OSP will include the level of security and operational permissions the staff will have, to properly maintain the system. This is done from a point of minimal access and builds, as needed. For example, you probably do not want a person routing calls with minimal training to have access to reboot servers or make system changes. However, you may very well want them to have visibility to the monitoring system so that they can properly route the ticket.

At a minimum, our Operational Support Plan should have the following elements:

- Monitoring Technology
- Service Desk Staffing and Hours of Operation (sometimes referred to as Service Desk or Helpdesk)
- Service Desk Escalation Procedures
- Trouble Ticket Tracking Software or Procedures
- Service Level Agreements for Support
- Troubleshooting Scripts (sometimes called Runbooks)
- System Administrators and System Responsibilities
- Engineering Staff
- Architectural Guidance and Review
- Change Review Procedures
- Data Continuity and Backup Plans
- Disaster Recovery Plans
- Business Continuity Procedures
- User Training and Enablement
- IT Knowledge Base

As you can likely tell, this could nearly be a book unto itself! It is very important, so I will attempt to identify the vital portions in this section with the hopes that you will be inspired to seek out additional guidance, to send me a note with $5000 and a gentle request to write said book, to create an online training course, or to train your team directly. (If you think I'm kidding, I'm not.)

Service Desk Operational Goals

A Service Desk (Helpdesk, Support Call Center, or whatever your organization likes to call it) is a group of people who respond to calls for help, proactive monitoring alerts, emails, or other ways of accomplishing the goal of keeping a Steady State and informing users. The goal of a Service Desk is very often to keep Subject Matter Experts, Administrators, and Engineers focused on their tasks as much as possible. Imagine how long the *Change* phase would take if you were interrupted every 15 minutes with someone needing a password reset. Maybe you can, if you are in that place right now. The solution is a well-trained Service Desk.

Everything—including trouble with expected performance, assistance or direction in accessing the system, determining a need for hardware replacement, or resetting a password—will typically come to the Service Desk. Service Desk staff will often create or route tickets for resolution internally or with the Administrators or Engineers (more on that momentarily, in the Escalation Procedures topic) and define priorities.

Although it often has a stigma associated with it in regards to competence, the goal of a Service Desk is typically having more people of a lower cost resolving or routing issues to those of higher cost or having other responsibilities. In many cases, the cost of a senior Engineer may be the same as three entry-level Service Desk workers. Therefore, it should be your goal to properly enable these workers, often working in call centers and often in 24-hour shifts. (We'll talk more about that in the Runbooks topic.)

Staffing a Service Desk is something that will need to be reviewed relatively frequently, especially in industries (such as retail or logistics) that experience surges in call volume that can be predicted, or when you have a major project change that will generate more calls (such as a hospital rolling out a new EMR system). How many staff and their level of experience will largely need to be determined according to the business's Success Criteria and just how much can be spent on staffing and systems to meet that criteria.

This can be difficult. For example, if the business leadership determines that user troubleshooting is essentially a best effort and that a one- to two-day response is practical because of productivity, they may only staff a few people in the Service Desk. You might be saying to yourself, *I know that will never happen. My leadership will want a 15-minute or less response but want me to do it as the Engineer, simply because they don't have the money to get a service desk going.* Here's the reality: they are spending the money, just not how they wanted to spend it! If the business has not properly defined the Service Desk staffing and is passing the burden on to the wrong people, you owe it to yourself (and them) to point out the issue. In most cases I have found, leadership is simply unaware of the problem. They aren't trying to be mean but they will try to expect the most of their staff, especially those who are compensated well.

Service Desk Structure

Here is a typical structure for a Service Desk that works with various combinations of roles. Note that a person on staff may spend a percentage of their time in one area, but also be expected to do others, as appropriate. For example, lower-volume Service Desks may often combine the Level 1 and 2 roles.

Here's how a typical Service Desk may be structured:

Level 1 Support Tech—First Response, Triage, Call Routing, and Priority Assignment. Observe and Report, unless directed by management to deal with issues not requiring troubleshooting, such as password management. Documents observations of the initial call.

Level 2 Support Tech—Basic Troubleshooting and Escalation Management. May also be asked to track the life cycle of a ticket and be assigned ownership of the Service Level Agreement (SLA). Will use a Runbook when possible and document the issue and steps taken for resolution or troubleshooting.

Level 2.5/3 Subject Matter Expert (SME)—Point of immediate response for known issues within a technology space. There may be multiple SMEs on a Service Desk, each with training or experience in their particular area of focus. You may have a Backup/Restore SME, an ERP SME team, or any other combination thereof. SMEs are often tasked with upkeep of Knowledge Base and Runbooks.

Shift Manager—Manages each shift in the call center or virtual call center to ensure calls and tickets are meeting SLAs. Note this may be a point of customer service or technical escalation as well.

Service Desk Manager or Director—Overall management of Service Desk staff. Will assign staff to shifts, tasks, and support levels. Generates and consumes reports to determine trends, effectiveness in points of escalation, and other issues. Coordinates with Administration and Engineering teams to report trends and will participate in Change Management meetings. Will communicate known issues and Change window impacts to all Service Desk levels. Will typically be involved in other management but not typically part of the technical escalation team for larger Service Desks.

Staff turnover, either by seasonal work or entry level employees, is typical of a Service Desk and the appropriate expectations and security measures should be in place. However, I believe that the strongest and most successful Administrators and Engineers I have seen in my travels have been those that started as Level 1 and 2 Support Technicians and "worked their way up" in the company by adding value along the way. The more you can inspire this by having success levels, goals, and rewards in your Service Desk, the better off your company will be. Plus, HR will likely thank you.

Service Level Agreements

Service Level Agreements (SLAs) are mutually agreed-upon expectations or timelines during which a problem is either resolved, updated, or moved to a different team. SLAs are typically determined by management or agreed upon during negotiations when they are a part of a larger services contract. I have encountered several iterations of this but the most common criteria for SLAs will typically contain:

- **Severity**—A measure of the priority of processing, based on the potential impact. For example, a Severity 1 (Sev1) incident may be serious enough to impact physical safety, have severe financial penalties, or be judged to impact enough people's ability to work to merit that all other priorities and processes be placed on hold. A complaint or feature request that does not have a deadline or ability to be quickly resolved may be a Severity 4 or 5.
- **Resolution Time Frame**—This refers to an agreement of when the ticket is either resolved or adjusted. This can vary, depending on the support given, but generally these guidelines give a busy staff member direction for which issues to work on first, based on their time "in the queue."
- **Status Reporting**—Simply put, this is the defined frequency with which official updates are given to the ticket owner, management, and/or issuer. These status updates largely determine if additional staff, vendor involvement, or management involvement is required. Because of this, it is not uncommon to see reprimands or penalties when the defined status window is not followed.

Service Level Agreements can be made using tables, with each of these three components as the headers or titles of each column. Because this format is not always ideal to portray these tables, as part of the bonuses of this book, we will include both PDF and Excel formats of all of the tables mentioned here.

Head to https://it.justdothis.net/mybonus to register.

Here is a sample Service Level Agreement structure:

Severity 1: Impacts a wider audience, and impacts safety or revenue in such a way that merits an "all hands" attention and suspension of other priorities and duties until resolved.

- **Resolution Time Frame**: Immediate to 1 hour, typically. This is largely subjective but the implication is that for some incidents of this type, there could be penalties associated for not resolving a revenue-impacting issue before the given timeline.
- **Status Reporting**: Between 5 and 15 minutes, depending on the incident response required. At the beginning of the incident, it will be determined if Leadership should be included and how often they will be updated.

Severity 2: Wider impact to productivity (for example, portions of a team or a whole team unable to work) or smaller impacts to revenue.

- **Resolution Time Frame:** From 15 min to 2 hours. The idea is that, save a Sev1 incident, this will be the team's top priority to resolve.
- **Status Reporting:** Between 15 and 30 min. Reporting is typically limited to IT leadership and the leadership of the team being impacted.

Severity 3: Impacts an individual or a smaller portion of a team where a workaround is not possible or production timelines, revenue, or data safety is impacted. Many teams will consider larger impacted teams where a workaround is possible as a Sev3, due to the lost productivity (for example, an EMR system's billing module is down but paper recording is possible).

- **Resolution Time Frame:** From 1 to 4 hours. This is the "bleeding but not life-threatening" type of incident. Issues of this type usually carry high visibility and a perception of high importance, but ultimately, must be preempted by higher priority impacts. Some teams will not distinguish resolution priority between Sev2 yet consider this a 24-48 hour resolution.

- **Status Reporting:** Typically every hour, unless an exception is made (for example, a vendor must be contacted or the change required for a resolution cannot be done during standard hours).

Severity 4: Impacts an individual who cannot work, or impacts a team in a way that is inconvenient or a low-revenue impacting way.

- **Resolution Time Frame:** From 24 to 72 hours. It is not uncommon to hear complaints that the issue "should not take that long to resolve," but if the team is working on a higher priority incident, they should not stop what they are doing to resolve this. This is why the resolution time frame is longer—not for the complexity but simply because priorities exist to protect the organization as a whole.
- **Status Reporting:** Between daily and upon completion are typical of this priority, depending on the organization's ability to automate or cope with volume.

Severity 5: Annoyances, feature requests, and other non-time-sensitive critical responses.

- **Resolution Time Frame:** None. At times, the feature request may be sent to developers, Application Owners, or the Architects. A resolution is not absolutely required for this unless a specific project is started as a result.
- **Status Reporting:** Ideally, this will be negotiated per request.

Ultimately, management and the IT teams should occasionally meet so they can review if the priorities set are working for the best interests of the company. Perhaps equally important is the consideration of having systems of tracking and automation, to make the process run properly. I have encountered a few companies that review this bi-annually, to achieve the right balance between frequency of changing systems and management satisfaction. Some organizations may go longer, and that may be acceptable.

The key here is making sure that everyone is in agreement and that the system is not abused by those with a kind of political influence within the company (for example, IT is shielded by the C-Level Executives from the Accounting department, who insist that the payroll system being slow is a Sev1 or Sev2 incident, therefore preempting an outage for the entire Medical Records department, who cannot work at all).

Many organizations may also appoint Incident Response Teams to manage Status reporting and then to evaluate the priority of tickets both in-flight and after the fact, to track their effectiveness.

Again, an entire book could be written about all of this!

Suffice it to say, someone ultimately needs to be responsible to ensure this is all working, even if that role is shared or rotates between management.

Escalation Procedures

An important element to any Service Desk is to determine how Escalation will happen. Regardless of the *how*, the methods must match the business's Success Criteria. I thought I'd get you started with some models I've most commonly seen.

Rapid Response Resolution. In high-volume call centers, often the best path is to keep a user on the phone and route the call to those that can help. This style is typically based on tight Service Level Agreement levels where each level of the Service Desk has a specific time to work on the problem before escalating to another level.

First Call Resolution. This method sets the goal of the first person to receive the call or trouble ticket to be the one solving the issue. While it is generally more expensive, this often results in a better user satisfaction because there are fewer people the user has to speak to in order to get a resolution. Some considerations for this type of Escalation method include:

- o First Responders must be well-trained.
- o First Responders must have a stronger sense of ownership. They will need to own the resolution of the ticket by reaching out to higher levels of support on the user's behalf. This often has a benefit of shielding the user from tickets being re-routed.
- o First Responders will need to be strong multi-taskers.
- o The business must be comfortable with a slightly longer time to resolution for each ticket.

Escalate to Resolution. Similar to First Call Resolution, this method allows Level 1 Support Techs to route a ticket to a Level 2 staff member who will ultimately take ownership of the call by any method to resolve with Administrators or involve Engineers. It is not uncommon for this style of ticket to call a user back rather than keep them on the line. At times, the Engineers will be tasked with contacting the users "owning" the ticket, but I will caution against using this practice because most Engineers I have met will immediately switch back to what they are doing and are much less likely to close a ticket or follow up. This is not a failing, but rather an appropriate focus on valuable tasks they need to be doing. The wise IT Manager will recognize this and have the Tier 2 person retain ownership of ensuring the ticket is closed properly when they get communication from the Engineer about it. Some will be encouraged to stay on the phone with the user and the Engineer to learn, which is a good practice, as I mentioned earlier.

Managed Resolution. At times, your organization may have specific owners for each account that are called in during daytime hours to manage a resolution response. Often called Technical Resource Managers or Technical Resolution Managers (TRMs), these specialized staff will work tenaciously to resolve a ticket. TRMs often have a proactive component that differentiates them from the rest of the service desk: they will often proactively call user groups or companies to let them know of new features, changes, or tips towards self-help.

I previously mentioned Service Desk roles, but I wanted to outline a full-tiered support system that has a time-tested success I believe merits consideration for any sized team.

Support Escalation Tiers

Tier 1

- **Members**: Level 1 and 2 Service Desk, Levels 2.5 and 3 Subject Matter Experts
- **Responsibilities**: This team is simply responsible for maintaining user satisfaction and making sure the business goals are being met. They are made aware of incidents and track them to a resolution.

Tier 2

- **Members**: System Administrators, Senior System Administrators, Application Owners
- **Responsibilities**: This team is responsible for the daily "care and feeding" of the systems, making sure that regular maintenance tasks are followed, and ensuring testing is completed properly. During escalation, they are typically contacted to conduct tasks that require experience or skillsets beyond what Tier 1 support staff are (or should be) entrusted.

Tier 3

- **Members**: System Engineers, Senior System Engineers
- **Responsibilities**: This team is responsible for changes made to the environment. They are also frequently responsible, along with the vendors and the Architects, for the development of new solutions as well as testing new configurations and upgrades. Escalations should be limited to highly unusual problems that require outside escalation (vendor support) that may involve making system changes.

Tier 4

- **Members**: Architects, Technical Directors, VP of Technology, Chief Technical Officer
- **Responsibilities**: These people are asked with ensuring that the business objectives are being met

by the technology. In terms of escalation, they are typically involved when a need falls outside of the solution or if the issue falls under multiple focus areas (for example, outages involving a network that effect storage).

Runbooks

The enablement of self-service for users, whenever possible, or the ability to automate trouble ticket creation often help to reduce Service Desk costs. However, since many users still need help beyond these first steps, you will need to properly enable your Service Desk workers. This is especially true of those answering calls. They are likely not very experienced but will still need to know how to escalate the issue properly. This is typically done by way of a Runbook.

Using Runbooks is often a good way of encouraging a process is followed (instead of relying on your own memory, experience, or "knowledge" to maintain a system). You have a process you are running though for a reason. In this case, the reason is that you need certain information or processes to be followed, to ensure nothing is missed prior to escalation. The reality is that Tier 1 people are typically paid much less than Tier 2 and certainly less than your full-time Engineers and Architects. Therefore, the more you can maximize their effectiveness, you can achieve two things:

- **A higher return on investment for hours spent.** The time spent by these individuals will have a higher value coefficient than if the same tasks were performed by Administrators or Engineers who have education and skillsets beyond what is required. You are taking time away from work Tiers 2 and 3 could be doing, if you have them working on these standardized steps.

- **Helpdesk staff gain experience in a safe way that builds confidence and increases the chances of them working towards a promotion.** Someone in a service capacity who is working towards a promotion is motivated to gain positive feedback results from those they serve. I shouldn't have to say that when I call into a Service Desk, I would always want someone motivated to make me happy rather than being merely a worker drone.

While there is no established standard for Runbooks, you essentially need them to accomplish a few key things:

- Guide the Service Desk worker in a standard series of questions or data to gather, in order to either resolve the ticket or escalate to the proper Subject Matter Expert (SME).
- Identify if the situation is related to larger outages and known issues with an established course of action (outages very often carry the course of action of simply informing the user, and filing the ticket under the known issue, for tracking and/or follow-up).
- Gather and document data involved in the call. Even when not relevant to the immediate circumstance, an analysis of some of this data can often prove very effective in identifying bad processes, trends, and workarounds that can be proactively dealt with.
- Guide the Service Desk worker in trying commonly repeated troubleshooting steps first (for example, a self-service password reset), and then, if possible, a guided resolution in a standardized fashion, typically instructing the user what to look for or do. Runbooks will usually have pictures or additional information so

the Service Desk staff member has more confidence in describing what should be seen.

- Have an identified path to troubleshooting and escalation for common issues as well as what to do in uncommon issues, when the steps identified do not work.
- Set expectations for support and escalation with both the Service Desk and the user calling for support. Depending on the organization, this may mean setting priorities pre-defined in the Runbook for the ticket (for example, keeping a password reset away from being a critical priority unless it happens to be a production database that impacts millions in revenue). The user calling may be told what to expect next and when to expect it. For example, if the Service Level Agreement (SLA) for a Severity 3 incident is 48 hours, the user is informed that they will be contacted within 48 hours with an update.

As far as who is responsible for creating Runbooks, this is actually where I see the largest risks in most organizations because they feel the people who created the solution should automatically be the ones to establish the Service Desk procedures. While I believe these people should likely have input, I would disagree that they be solely responsible for establishing such procedures.

For example, hiring a consulting organization to conduct a design and implementation of a project could include in the scope guiding Success Criteria for Runbooks (identifying commonly known issues and how they are resolved or identifying points of concern during an implementation that could not be resolved) but having these individuals tasked with the actual creation of support Runbooks is rarely successful.

Ultimately, those involved in the support process who can identify repetitive tasks or procedures that are done often by individual tickets escalated "too high" should be writing the procedures (or at least responsible to see they are done). Regular reviews of trouble tickets to identify trends which do not fall into the current Runbooks is important.

The most common updates to Runbooks come from the Testing procedure within the *Change* phase. During Testing and the Pilot Rollout, the following should be identified, documented, and evaluated as they are encountered, because this will set them up for being included in the Runbooks or dismissed as uncommon:

- Administrative Prerequisites (account access, groups, etc.)
- Client Prerequisites (for example, software versions that must be present)
- User Prerequisites (education, documentation, or signed pledges)
- Procedures for Access
- Procedures for Work
- Issues encountered (common issues noted with how many times per population they occurred)
- Resolution Steps (for any issues that are encountered)
- Expected/Observed Performance Baselines

The most successful support organizations I have encountered have regular reviews of these procedures. Those that skip this step repeatedly because they "don't have time" or feel it is not valuable very often find their staff and users to be frustrated when the Service Desk individuals automatically escalate each ticket. In my experience, this is especially true of any application access technology.

In my previous work, the phrase "Mind your ABCs: Always Blame Citrix" was all too common and preventable by simply empowering the staff encountering the issues to identify common issues before escalating. I typically found complaints among Citrix Engineers in these organizations that if the word "Citrix" appeared anywhere in the ticket, it was passed along to them with no attempt to resolve. In one case, four Service Desk staff, each making about $35,000 a year, were mostly idle through the day, while the Citrix Engineer making $85,000 a year was working overtime—and was ready to quit because they were spending most of their time using the same tools available to the Service Desk staff for tasks like password and profile resets! In fact, in 2017, in over 30% of the consulting projects with which I was involved, the reason I was brought in was because the Engineer had quit and they had no coverage for the tasks that were required. To my lack of surprise, my initial evaluation found operational gaps that caused their Engineer to quit because they were "stuck" doing repetitive tasks that management refused to recognize.

So, in my mind, it is simple enough: take the time to enable those in your Service Desk to be effective or you will ultimately end up hiring someone like me—for $350 an hour—to correct the issues (or worse, plug the holes). It's your decision!

Administration and Engineering

I will begin with a caveat: I have seen a LOT of variance in how these job roles are either defined or more appropriately, how they work in practice. That said, there are certain job functions that I feel any services organization must accomplish, which are woven into the phases of the methodology. I will outline the two most successful methods I have seen and then outline a typical layout for each, based largely on either the work complexity or the company size.

The first and most common model doesn't really have a name because it is based on maintaining a Steady State and the roles do not usually change. In this context, Administrators and Engineers have well-defined job descriptions (at least, they hopefully do) that allow everyone to know at a simple glance where a task's resolution will be.

Here is the structure I have seen most successful, followed by what I see at larger organizations.

First, the Tier 2 Administration team that is crucial to the Maintain phase.

Systems Administrator

- Low-impact Administration duties, including adding users to system access groups or setting up user accounts and profiles.
- Monitoring systems and proactively or reactively alerting Senior Administrators or Engineers of issues that cannot be resolved with their level of permissions or scope of operations.
- Typically the first point of escalation from Tier 1 teams.
- Some Administrators are tasked with ensuring backups are functioning and performing restore operations.

- Often tasked with updating certain lower-impact systems and users' systems.

Senior Systems Administrator or Systems Administrator II/III

- Often a multi-practice Administrator with several years of experience who has a wider understanding of the IT systems as a whole and can inform or guide Junior Administrators and Engineers as to the "lay of the land" or how things are typically done.
- Tasked with updating servers and granting access to privileged areas or data.
- Senior Administrators are more seasoned to handle higher-impact (therefore, higher stress) tasks.
- They may be responsible for maintaining physical or virtual environments or specific software not otherwise defined (see "Larger Organizations" later in this section).

Lead Administrator/Systems Manager

- In larger organizations, the Lead or Manager is generally responsible for the division of tasks and priorities.
- They are the point of coordination with Service Desk, Engineering, Architectural, and Management teams.
- They typically spend as much time in meetings as at their computer and are often on a Management track.

Now, we move on to the Tier 3 individuals—the Engineers who typically spend most of their time in the *Change* phase.

Systems Engineer

- Task-oriented changes that require a more experienced person or are happening outside of regular production (for example, testing or validation environments).
- Often tasked with specific change tasks on a regular basis that are not appropriate for Administrators to complete or require higher certifications per company or vendor restrictions.
- Engineers tend to be "heads down," interfacing a lot less with users but often part of a larger team.
- Sometimes the 3rd major point of escalation for issues.

Senior Systems Engineer or Systems Engineer II/III

- Charged with interpreting the design of the Architects or Consultants and either enacting the change or delegating the changes to a Systems Engineer.
- Very often assume responsibilities for minor design work (which are, unfortunately, not as often documented).
- Test version changes for impacts to production, procedural changes, or notes for senior staff to consider in decision making.
- Often charged with evaluating new products or procedures and documenting the results.

Lead Engineer, Project Lead, Principal Engineer

- Much like the Administration counterpart, the Lead Engineer is responsible for ensuring work is done at the right times. They are the people responsible for reporting and documentation to management and the other teams.

- A good Lead candidate will not only be a good leader but will be more personable than the "typical" Engineer. (The honest truth is that not every person suited for engineering work is suited for leadership, due to the nature of the work. Engineers should not have leadership in their career path unless they really desire it.)
- Project Management skills and the ability to estimate time for tasks to be completed is important for these people to have.

Larger Organizations

In larger organizations, you may find roles are further split by the technology focus. For example, you may have Systems Administrators who are focused completely on maintaining the hypervisor and maintaining the physical servers, specifically. These people are often, unsurprisingly, called Server Administrators. In my travels, I have most frequently found these individuals operating at a Senior level.

Likewise, you may find this to be true for Engineers who are tasked with a certain specialty at least 50% of their time. I have found this to be true in organizations with systems that require maintenance and changes frequently but not constantly. The Engineer is then tasked with other items to make up for the time, even though their title may reflect their focus and seniority. For example, a Citrix-focused Engineer who is typically responsible for building primary systems and delegating work to others may be classified as a Senior Citrix Engineer or a Systems Engineer II—and everyone knows they are the main person for Citrix. This person may also have the responsibility of building or maintaining the hypervisor cluster in their scope.

An additional note here: the Lead or Principal Engineer should be ready for overlapping time demands and have the ability to effectively translate technical jargon to management. Think of *Office Space* where a person's entire job is to interface with Engineers and customers. While it is seen as wasteful, I disagree. Put bluntly, I've known too many Engineers. (And at this moment, I'm wondering if I can do a "winky face" emoji.)

Consulting

Those service organizations that perform work externally (i.e., Consulting or Service Providers) may also label the roles differently. Engineers are often called Consultants, with distinguishing levels, such as Staff, Junior, Senior, Lead, or Principal. As surely you've come to expect by this point, the line with Consulting is not one I intend to cross in this section because it implies drastically different responsibilities and methods that need to be in a dedicated work, even though our Methodology absolutely still applies.

Architects and Leadership

I have been known to say that "Architect" is a role people want to be in until they get there. This is because very few technical-minded people are truly equipped to deal with the internal political struggles, balancing expectations and agendas, or the team dynamics often associated with such a role. And – many dislike writing which is what an Architect spends a large amount of their time doing.

Now that I've properly gotten your attention, an Architect is typically a person with years of experience. IMPORTANT: I am specifically saying "experience," NOT "certification" here, for a very good reason. The Architect should be someone with a high amount of business as well as technical acumen. The primary role of the Architect is to take the requirements of the business and translate that into what is possible in the technical realm. This is quite simply not something that can be taught in a certification program or, unfortunately, a book.

The Architect primarily spends their time in the *Understand* and *Plan* phases. In fact, I wrote this book with a lot of content in those phases so that you will be able to gain enough meaningful experience to work up to an Architect role or, at least, decide if that is the path you'd like to pursue. That is also why I have seen Senior System Administrators essentially skip the Engineer role completely and jump to Architect. They do so by participating in the *Understand* and *Plan* phases so frequently that they were given tasks to do. They performed them well while showing the appropriate "connection" with the users' and business's needs.

The key skillsets of the Architect include (but are certainly not limited to):

Technical Authority in a Focus Area. It isn't just about "paying your dues" and being in a position for five years and being promoted to Architect. At least, it shouldn't be. The Architect needs to be able to speak definitively on Leading Practices for their technology. They must be aware of what works for other companies and adapt those strategies for what is relevant in their environment. This may require occasionally conducting or observing Engineering tasks. More often, it means spending about 20% of their time in study.

Extreme Ownership. The concept put forward in Jocko Willink's book, *Extreme Ownership*, isn't just simply a willingness to fall on one's sword if something goes wrong. The biggest reason that things go wrong in IT projects is actually a failure to identify the requirements and capabilities of the design, which falls on the Architect and is why this role tends to be more highly compensated. A person that is quick to shift the blame is not a good fit as an Architect. For example, look at the concept of taking responsibility for things that went wrong. What should be done is this: go through post-mortem exercises with the team and find ways to improve. Face those fears.

If a company has grown to the size that requires a dedicated Architect, they have likely also embraced the concept that mistakes happen and will hire people willing to take ownership, learn, and adapt, without falling into despair. If you aren't working a company with that kind of belief, it may be time for a change.

Leader among Leaders. The Architect needs to be someone who can work quite often with other Architects, Directors, and those in CTO and CIO positions. This means a unique quality and personality is required. Pride and politics will get you only so far when it comes to working with others, and believe me that if you want to be working in an organization where you can make an impact, you will absolutely need to work well with others.

This means learning to compromise when it is needed, and also to make a strong case for your group's needs, when they arise. The more you can learn about your fellow leaders and be able to meet their needs in what you do, the more successful you will be in your area of focus.

Demonstrated Ability to Guide Tech Teams. This is often confused with management. "Guide" does not mean you are responsible for the resulting actions. Please hear and feel that. GUIDE does not mean the Architect is responsible for the resulting ACTIONS taken by the staff. It takes a special personality to perform this role. By focusing on Design Intention, not end product, the Architect can adapt and inform their team and other teams about Leading Practices.

Community Leader. I personally think that every technical Architect should be an active part of the local or online tech communities that align with their role. They should be intentional about building others up. The Architect will also be involved in hiring decisions, most of the time.

Heart of a Teacher. A successful architect is skilled in explaining WHY and HOW a technology works to their team and the teams around them. Remember that an Architect tends to be focused on a technology area, which means they will frequently be working with other Architects or Consultants. Take the time to explain WHY something needs to be a certain way and you'll get things done faster with better relationships resulting from those interactions. I have very often observed Storage Architects at odds with Database Architects because the Database Architect simply makes demands without explaining why or collaborating on how to do it better. (One incident I can recall involved two Architects who literally sat next to each other and never really explained why their systems had the requirements and limitations they did. When I came in as a consultant and asked what storage systems were present, I found that they had an underutilized the fast storage system able to house their VDI profiles more effectively. If the Virtualization Architect had simply explained the need to the Storage Architect and taught him a little bit about why it was important, she would have had a much better performing environment.) The same can be said for Engineers and Administrators. Those who understand WHY make smarter decisions. Those who better understand the technology escalate less often. Management who understand the technical requirements in terms of business needs approve budgets, so teach first!

Be a Generalist Know-It-All. The contradiction here is on purpose. The Architect needs to know about their technology and be an authority, but quite often they are called on to be highly knowledgeable in other areas as well. Some Architects are, in fact, the only person at that level in their IT department. This means they must be a point of authority for EVERY technical area. To do this, you must have the curiosity of a generalist with the confidence of a leader. So, again keep in mind that pride will get you only so far—but your organization needs you to be appropriately confident about what the technologies do and how they interact, and most importantly, to be able to identify risks.

Some organizations may make the mistake of requiring someone to be a generalist Architect who also performs

Engineering and Administration tasks for other technical areas. I have been seeing a trend toward "not keeping all the eggs in one basket" mindsets causing this kind of structure to happen. The mistake here, and feel free to share this paragraph with anyone who needs to hear it, is that if you require those that should be thought leaders to spend too much of their time *not* innovating, they won't innovate when they need to. Why? Because even if they magically manage to find the time, they will not want to implement something new that they have to learn, build, and support. Your company will stagnate from a technology perspective in these scenarios, and more importantly, the Senior Tech/Architect will rarely stay. There are other jobs out there. Of course, as a consultant, I realize that I'm making a strong case for hiring a consulting company to do the thought leadership, so let's dive into that a bit, shall we?

Contractors and Consultants

"Don't own stuff."

—Common saying among CIOs

A growing trend in IT technology—and arguably, another reason I wrote this book from the perspective that I have—is that many organizations would prefer not to *Maintain* "stuff." Stuff could mean servers, buildings or even staff. When you think about this, it makes sense. After all, when it comes to employees, some of the largest costs are not always the salaries. The cost of benefits, taxes, and required administration overhead for staff positions are so high that having full-time employees is often hard to justify.

Many companies are, therefore, finding ways to identify needs for which they can keep a full-time staff when they need to for specific functions. For example, in most cases, it makes sense to keep a sense of ownership in administration. However, outsourcing the Service Desk will often make sense, from a cost perspective, because needs fluctuate. But in terms of design and build activities, quite often, it also makes sense to only have people involved for said projects on a temporary basis. By hiring contractors and consultants to perform these projects, the company does not have to invest in training or in keeping a "bench" of expertise not being fully utilized.

This is why I continue to predict that the largest growing opportunities to specialize in a technology and get paid a lot more will come from being a Contractor or a Consultant. There is much to be said here, and because of this, I also conduct separate trainings and books on that topic.

To stay up to date on this information, register for the bonuses and remain part of the mailing list at https://it.justdothis.net/mybonus

For now, here are some things you should know:

A Contract Engineer or a Consultant both operate within a Scope of Work. This defines the specific needs and deliverables of the project and sometimes the length of time of the engagement.

A Contractor will typically work on an as-needed or temporary basis. This can happen independently, although the relationship is more frequently brokered with another company who will be responsible for the invoicing and insurance for the project. Most clients prefer to work this way because they can come back to the contracting company for other projects as other needs arise, and keeping one Accounts Payable relationship is a lot easier than keeping many.

Consultants, in the same way, can be hired to provide expertise and design documents which can be handed off to other Contract Engineers to perform. Becoming a trusted Consultant typically takes years of experience but I can tell you firsthand that the rewards are numerous, especially when working independently because of one word: "No." However, many consultants actually work for a company—most often a reseller. Other companies, such as Citrix Consulting, operate a blend of full-time and Contract Consultants and Engineers.

One area that can be a little confusing at times is when a company will refer to their internal staff as "consultants" even though they do not operate within Scopes of Work. They typically do this when they want a person to be a thought leader but not be considered an Architect. I wish I could say I understood why I see this, but there are no specific standards for these types of positions. So, that means you must read job descriptions carefully, even if you will be working for a company that services other companies!

Chapter 27: Restarting the Cycle

Before we close, I wanted to remind you of a key concept: This is an ITERATIVE Methodology. It is crucially important to return to *Understand*, continue to *Plan*, implement a *Change*, and move back to *Maintain* on a regular basis.

If you are wondering when to restart the cycle, I have already mentioned several examples but here are some additional scenarios:

- Regular Operating System Updates
- Regular or Irregular Software Updates
- Release of New Features
- Software, OS, or Hardware going End of Life (EOL)
- New Feature Requirements
- New Hardware Requirements (i.e., storage)

What people often forget to include in the business of either keeping a Steady State (*Maintain*) or when they are overwhelmed with so many of the circumstances (like those just listed) is that it is healthy to re-introduce the cycle.

What I specifically suggest is making a schedule of when to create a new cycle for every major IT system.

Here is what you should do (or consider doing) monthly, quarterly, and annually:

Monthly

Performance Review. Compare current system's performance against baselines to note improvements or declines. It is good to note what is working and what is not, and to suggest adjustments during the next Change window, if appropriate (or, if required, the next major project). Use your monitoring tools or other observations (for example, load or login time averages) to compare trends.

Service Desk Ticket Trends. One of the more commonly overlooked indicators of both trouble and success that you can use—especially for your management—is taking the time to note trends in Service Desk tickets. Beyond metrics, remember that these are indicators of both pain and relief (in the downward trends, they indicate relieved pain) that users are facing... and will be telling your management about. If you note them proactively, you can act before a problem becomes bigger, note issues in other systems that can affect yours, and have evidence of your wins to present to management.

Quarterly

Workload Growth Analysis. Do the capabilities of the system meet the current needs *and* the predicted needs for the next quarter? This is an area of risk that is often assumed to be covered but in actuality is not. Countless times, I have encountered systems overwhelmed because three years prior, they were designed for a certain load, only to have an unannounced merger or acquisition drastically increase the load without proper hardware being added. By reviewing this quarterly, you are more likely to be able to identify these trends, respond within the next budget cycle, and/or let leadership know of the problem to come. For example, you will have time to alert leadership of any shortcomings regarding most major additions of staff or workload, even if you feel the system can handle the load in its current state. Either way, it is important to review this regularly.

Staff Utilization Analysis. Although this is much more of a management-level task, it is important to recognize when your IT staff is at the proper utilization level. Are projects being completed on time? Are SLAs being met? By reviewing these quarterly, much as with workloads, you can more proactively inform leadership of capabilities and requirements before they become a problem. Do not make management's judgments for them. Very often, I have encountered teams who did not perform this analysis simply because they felt it would be "a waste of time because management wouldn't approve more staff acquisitions anyway." Don't do their job for them. Instead, make it easier for them to make decisions based on observations and facts, and your department will be the one run most effectively. I'm pleased to say I have encountered teams that have more than enough staff to handle issues that come up or more intensive projects— and this Staff Utilization Analysis is a hallmark of their success.

Service Desk Trends Analysis. Just as with the monthly review, this review notes at a larger scale what tasks occupy most of the staff's time so that in concert with the Staff Utilization, a report can be generated as to how effectively IT is meeting the needs of users. Have tickets increased or decreased? Report this data to leadership quarterly, so that they can make decisions accordingly.

Leading Practices Health Check. IT systems are very often the largest persistent investment a company makes. So, I tend to think of them like a supercar. While we love old sports cars, those we call supercars are on a quest for constant improvement and being the absolute best they can be. What worked last year won't work this year. Some car manufacturers (such as Tesla) have gone so far as to make improvements on a much more regular basis. This is not done by hope, dreams, or some artistic fantasy. It is done by a regular and purposeful review of what is working and what is not, even for their competition. Seriously! In researching this concept, I found that almost as many supercars are sold to other car companies as to individuals. These companies take them apart, learn what they can, and see how they can do it even better. Twenty years ago, I could not conceive of owning a car that would achieve going from zero to 60 MPH in under five seconds. I recently bought one, for under $35,000. Without this constant will to improve and do so reliably, it would not be this way. It also would not be that modern supercars (as of this writing) can achieve this same feat in two seconds.

Another thing here: there have been many high-performing vehicles that proved ineffective over time. Why? They did not have hoods! Why would you buy a car with no hood? How could it ever be worked on or improved? In the same manner, it is very important that you not assume that you will be able, on your own, to determine the best way your systems can be configured. You must seek outside opinions. So, I recommend a regular review of what other companies and even the vendor consider current Leading Practices for the specific technology under your charge. In most cases, you won't be terribly concerned about the other systems. Be honest about where your systems are and consider a "Top 10" style list of items to always cover to ensure your configuration makes sense against what others are doing.

Annually

Infrastructure Analysis. While a Health Check covers Leading Practices in general, an overall Infrastructure Analysis goes beyond this to review business goals, identify changes therein, and make sure that they are being met by the IT systems. You'll want to produce a document (covered in the Documentation section) that covers the business goals and objectives and details how all systems are working together to achieve this in the best way possible. If there are variances or risks present, the next step in the process will be to plan a new project, to resolve these gaps. A project of this type should also make sure the proper licenses are in place.

I typically recommend this yearly task be performed by an outside consultant. Why? Primarily because when you work close to a situation, it is more difficult to see the flaws. This is something I often remind my coaching clients of. An outside perspective helps you see around corners, so to speak. Also, hiring a consultant or company gives you the added advantage of professionalism. A project of this type will typically take between one and four weeks to complete. By the way, if this coaching or consulting describes something you would be interested in pursuing, please email me at support@thrive-IT.com and we will set up a complementary consultation. It is my gift to you.

Design Review. While often combined with the Infrastructure Analysis, the Design Review focuses on a specific technology component to ensure the design is meeting the business objectives. In addition, this is the time to look ahead and make sure there is a plan to address hardware going End of Life (EOL), the proper use of Cloud technologies, and any EOL for operating systems and key software. A design should be made after this review that addresses gaps and risks, and may include spending over the coming year. For that reason, I find that this project is best done before the annual budget review is to be conducted.

Operational Support Plan Review. While the Design Review may indicate a need for supporting technology changes, it will also note how effective the Support Team has been over the past year. In addition to this, include a full review of support tickets, response times, and SLAs that may need adjustment. I typically recommend this review be conducted just following the Design Review's *Plan* phase, to indicate changes to the staff structure or procedures to better support the technology. Changes to Runbooks or other documentation for the support team are also appropriate here.

If you follow the structure I have outlined, you will have no problem restarting the four-phased cycle of the Methodology: *Understand*, *Plan*, *Change*, and *Maintain*.

CONCLUSION

Chapter 28: Go Do This

As we wrap up, there are a few things I wanted to add. First – that in the process of writing this book I came to realize that I needed to change my entire business structure. Since starting the writing process to now typing these words, I have had the opportunity to meet many of you and hear your stories. You inspired me to help even more, to go even deeper. As of this writing – this is now my full-time purpose and I do very little consulting. I have even had the opportunity to bring others onboard in this mission which is amazing, scary, and exciting all at the same time!

My team and I at THRIVE-IT.com are continually working on a series of resources and bonus materials for you. I am confident these will add to your Mastery, Methodology, Mentoring, and Mindset to grow your professional and personal life. We are building a community for people just like you, who are dedicated to improving themselves, advancing their careers, and making an impact on the world. These resources are available at https://it.justdothis.net/mybonus.

Now, one final thing, before we part ways.

When I was sitting in my living room chair, reviewing the edits for this book project, I had all of the *Star Trek* movies playing on my TV, in the background. As I was approving the edits for this final section, the background show was the last *Star Trek* movie to be produced—*Beyond*. It can't be coincidence that as I was adding the last words of this very section, the phrase "To boldly go where no one has gone before" was sounding out in my home. I knew I could not simply end with an invitation to read more.

This is my desire for you: I want you to go BOLDLY, in confidence of a rare opportunity you have. Information Technology has done great things in the world and allowed countless innovations. It has created opportunities for humanity to move forward, despite our background, our race, our past, or the challenges with which we were born. You now have the opportunity to THRIVE—to use technology to solve problems, move forward, and boldly go into the future, with a vision. If there is to be a vision, may I submit to you my guiding vision—something that I shared with Mark Templeton years ago? I hope you will tear up the same way he did at hearing it, knowing that it is feasible:

"To leave the world better than I found it, and to equip and inspire others to do the same."

Go forth in the confidence that if your work is connected with a true reason, a WHY, your WHAT will have more purpose. And if that isn't true today, I know that using the strategies I've outlined here will help you move to where it is true—just as I have. You will find ways to go beyond just the technology itself and find the ways to equip people to work better, be better, and do better things. They are waiting out there, and you can be part of what helps them make a difference. You can help hundreds and thousands do more than what they could do on their own. We really are better together.

Just Do This!

What To Do Next

Was this book helpful? I'm going to assume so if you've read this far.

I have a favor to ask! I am self-published which means I have an advantage and a disadvantage. My advantage is that it is very easy to make corrections and new versions of this book pretty easily. My dis-advantage is that I don't have the marketing ability of large publishers. However – you can help me spread the word!

Here's what to do:

1) Go to Amazon, find Just Do THIS and leave an HONEST review. I thrive on honest feedback – it helps the next book be even better. Tell people about what you liked and didn't like. Your review (not just your rating) helps people decide if this is the right book for them to invest the time in reading!
2) Share it with a friend! You can always share either our blog at it.justdothis.net or methodologybook.com. Tell them to email me with your name as a referral; as I'm able I'll be sending out thank you messages and gifts for those that refer others!
3) If your team would benefit from a bulk order of this book – let me know! There's some great options.
4) Finally, if you would like me to come speak or teach at your event or offices, please reach out at support@thrive-it.com.

Thank you for taking this journey with me! If there's anything I can help you with in the figure, please let me know! But more importantly – if you have gotten a benefit from this book, resources, or courses – please send us an email! Remember – I'm on a mission to create $100,000,000 in raises, promotions, and new career opportunities and would love to know if you were one of them!

About the Author

D.J. Eshelman grew up in Colorado (U.S.A.), where he lived until 2017. He currently lives near Nashville, Tennessee.

After a successful 20-year career as a Citrix Engineer and Consultant, DJ is now dedicated to equipping the next generation of tech leaders through creating books and training material to equip professionals with what they need to be successful.

D.J. still occasionally serves as an independent consultant for Citrix Consulting and other organizations.

He has served companies of all sizes, including 20 of the Forbes Top 100 companies, and among others has these credits to his name:

- CCE-V (Citrix Certified Expert in Virtualization)
- CTA (Citrix Technology Advocate) 2017-2019
- CUGC (Citrix User Group Community) Leader— Founded Colorado Chapter in 2016. Co-Lead Nashville from 2017-2020.
- Creator of CTXPro.com
- THRIVE-IT.com—Creator & Lead Coach
- Resident Consultant with Citrix Consulting Services

D.J. is also a Life and Career Coach and is a firm believer that to be most successful, we must have the attitude of leaving the world better than we found it.

D.J.'s debut book was *Be A Citrix Hero*, which is available via Amazon and other select booksellers or at https://ctxpro.com/book

Acknowledgements

Finally, thank you to my book launch team who were helpful in proofreading, making suggestions, and sharing the book!

David Ward
Liang Cheng
Chris Daly
John Rohland
Mike Doolittle
Kevin Graham
Hector   Miguel
Guerrero
Harshit Shah
Manuel Winkel
Ashley Barlow
Fred Wix
Ray Anderson
Eric Nettles
Kevin Benson
Gareth Joyce
Pavan Ayyagari
Fabian Danner
Erkan Sefiloglu
Devjit Banerjee
Stephen Curley
Abraham Brown
Carl Behrent
Matt Hambley
Vijian Narayanasamy
Phil Bossman
Florin Robete
Jon Falgout
Sajith B
Jeff Pitsch
Muris Aljukic

Henry Heres
Jane Cassell
Benjamin Crill
David Sharp
Giovanni Svette
Bart Jacobs
Esther Barthel
Samuel Asiko
Nils Michel
Donald Wong
Kris Radke
Thomas Link
Steve Topazi
Parveen Singh
Gurprit Virdi
Travis Shackelford
Justin Lippman
Daniel Schmeltz
Thomas Freitag
Danish Haider
Adam Forsman
Jaco van Diggele
Vivek Sarmalkar
Jason Stacey
Scott Munger
Keegan Shearer
Deniz Coban
Mike Biracree
Anup Kumar
Alexandru Pastiu
Brandon Mitchell

Ahmed Adeoshun
Larry Henshaw
Bryan Dial
David Cobb
Ganesh Kumar
Kevin Goodman
Patrick Arnold
Michael Roberts
Kenneth Shaddle
Abraham Brown
Amanda Heiser
Craig Stones
Adil Nasser
Anup Kumar
Douglas DeCamp
Chris Schrameyer
Giri Sonty
Mark Gerk
Larry Henshaw
Paul Willett
Steve Wightman
Sergey Kluzner
Matt Hambley
Nick Hails
James Shields
Kumar Rajeev
Venkatesh Murali
Ken Avram
Brandon Brewer
Amjid Khan
Ray Reyes

Made in United States
Orlando, FL
08 March 2022

15544876R00154